D0146621

Science and Technology of Organic Farming

Science and Technology of Organic Farming

Allen V. Barker

CRC Press is an imprint of the
Taylor & Francis Group, an **informa** business

CRC Press
Taylor & Francis Group
6000 Broken Sound Parkway NW, Suite 300
Boca Raton, FL 33487-2742

Printed in the United States of America on acid-free paper
10 9 8 7 6 5 4 3 2 1

International Standard Book Number: 978-1-4398-1612-7 (Hardback)

Library of Congress Cataloging-in-Publication Data

Barker, Allen V., 1937-
Science and technology of organic farming / Allen V. Barker.
p. cm.
Includes bibliographical references and index.
ISBN 978-1-4398-1612-7 (hardcover : alk. paper)
1. Organic farming. 2. Agricultural innovations. 3. Agricultural chemistry. I. Title.

S605.5.B365 2010
631.5'84--dc22 2010009478

Visit the Taylor & Francis Web site at
http://www.taylorandfrancis.com

and the CRC Press Web site at
http://www.crcpress.com

Contents

Preface

Organic farming is said to be the original, mainstream form of agriculture. Before the development of synthesized fertilizers and pesticides, practices of crop rotation and fertilization with animal manures and legumes were the options available to most farmers to maintain crop productivity. All systems of organic farming avoid the use of synthetic fertilizers and pesticides. However, the principles involved in organic farming are in common with conventional farming—a term that is used to identify agriculture that is not organic—as both systems are managed intensively. Production in the wild or in abandoned systems is not an organic practice. Crops must have nutrients, and pests must be controlled. Many conventional systems have practices that are in common with organic farming. Hence, organic and conventional systems are not greatly different in technology and have the same scientific basis.

The science of organic farming has emerged as a combined study of soil fertility, plant pathology, entomology, and other biological and environmental sciences. Organic farming is a philosophy but also is a well-researched science, for which methods of application are presented in this practical text. The purpose of *The Science and Technology of Organic Farming* is to provide a current, readily available source of information on the scientific basis for organic farming and the technology to achieve adequate yields through plant nutrition and protection. Farmers should be able to dispel myths and philosophies that are not the substance of organic farming.

Chapter 1 provides an overview of organic farming as well as the processes and results of attempts to define and regulate organic farming. Organic gardening began unscientifically but has adopted traditional scientific methods in its present form. Soil fertility and plant nutrition are discussed in Chapters 2 and 3, with emphasis on the macronutrients nitrogen, phosphorus, potassium, calcium, magnesium, and sulfur and a group of plant micronutrients. These chapters are major components of the book because failure to provide adequate nutrients to crops is a limiting factor in organic agriculture. The chemistry of the plant, the soil, and the soil solution are explained in the chapters on soil fertility and plant nutrition. Guidelines, recommendations, and procedures for determining the best fertility recommendations for individual situations are presented.

Chapter 4 covers practices for management of farm manures for incorporation into land. The practice of returning manures to the soil improves fertility of the soil and has been researched as a viable system of effective soil management in all forms of agriculture. Chapter 5 discusses the purposes and the production and assessment of compost as a factor in maintaining soil fertility, and Chapter 6, on managing green manures, follows and completes the discussion of organic matter in soil fertility.

Chapter 7, which discusses liming, is presented to introduce the effects of soil acidity and its management on soil fertility. The acidity of agricultural soil affects the microbial activity and the availability of plant nutrients from organic and mineral matter and is a major factor limiting crop productivity in the world. Mulches and tillage are covered in detail in Chapters 8 and 9, respectively, followed by chapters on methods of organic weed control (Chapter 10), insect control (Chapter 11), and disease control (Chapter 12). Organic pest control is a substantial component of management of soil fertility and crop productivity in organic farming. Researchers are working to advance this segment of the field, and some developments are reported.

The book ends with discussions of companion planting in Chapter 13 and the storage of produce in Chapter 14. A glossary is included to assist readers in terminology used in the book. A bibliography is added to acquaint readers with some of the literature that was consulted in the preparation of this book and that readers may want to read for more information.

Organic farming offers a scientifically derived method of improving soil fertility and for increasing yields with limited chemical inputs. With the world population increasing rapidly, and projected to do so for some time, and with improved plant nutrition and crop protection as major factors increasing crop yields, use of the knowledge of organic farming to maximize agricultural yields and to sustain soil quality will grow in importance. Public interest in minimizing the use of chemical inputs in agriculture is increasing the demand for organically grown foods.

This text is intended to be a practical handbook of organic farming and a reference work for anyone interested in organic farming. It will also give information on how to assess and govern the nutritional status of crops and the fertility and condition of soil. It should be of use to farmers, agricultural advisers, soil scientists, and plant scientists.

The author gratefully acknowledges his daughter, Robin V. Barker, for providing the hand-drawn illustrations in the book.

Allen V. Barker

Author

Allen V. Barker, PhD is Professor of Plant, Soil, and Insect Sciences at the University of Massachusetts, Amherst. He has over 45 years of research and teaching experience in organic and conventional agriculture and has interests in plant nutrition and soil fertility. He is a retired farmer. He grew up on a crop and livestock farm in southern Illinois. He graduated in Agricultural Science from the University of Illinois and received master's and doctoral degrees in Agronomy from Cornell University. He regularly teaches Organic Farming and Gardening, Soil Fertility, and Plant Nutrition at the University of Massachusetts.

1 Definitions and Philosophies of Organic Farming

HISTORY AND BACKGROUND

The term *organic farming* was introduced into common usage around 1940, following farming movements that had begun in the 1920s and 1930s promoting the concept of management of a farm as a living unit or whole system. During the period of the 1920s and 1930s, Albert Howard in the United Kingdom, and based on his work in India, laid out the social and practical groundwork for the organic gardening movement. Rudolph Steiner, through his lectures and teaching beginning in 1924, laid the foundation for biodynamic agriculture, which created the first organic-like certification and labeling system. However, biodynamic agriculture differs from organic agriculture in that the biodynamic system has spiritual, mystical, and astrological guidelines. Lord Northbourne, an agronomist in England, in reference to farming, introduced the term *organic* to the world in 1940 in his book *Look to the Land*. J.I. Rodale introduced the organic movement to the United States about 70 years ago with publications that advocated for health through farming organically. During the 1940s, the Rodale Institute (United States), the Soil Association (United Kingdom), and Soil and Health (New Zealand) were founded as associations devoted to study and promotion of organic farming. Similar organizations arose in Germany and Japan at about the same time or just following the establishment of the institutions in the United States, United Kingdom, and New Zealand.

At the time of the early development of the organic movements, however, use of chemical fertilizers and pesticides was very modest compared to current practices; hence, the development of organic farming was related not only to the materials used for soil fertility and crop protection but also had a base of managing a farm as a system with integration of soils, crops, animals, and society. This concept of a systematic approach is fundamental in organic farming today and is expressed with opposition to production of genetically engineered crops and irradiation of foods, among other practices. The association of organic agriculture with environmental sustainability arose in the 1960s and 1970s and brought about changes in the politics and social elements of organic farming. The involvement of governmental agencies in organic farming increased markedly during the 1980s and later.

Until recently with the applications of legal restraints to organic farming, no universally accepted definition or identification of organic farming and gardening was developed. The difficulty of defining organic agriculture arose from multiple conceptions of the basic nature of the term *organic* among biologists, chemists, and practitioners. In some cases, terms such as *naturally grown*, *wild*, *biologically grown*, and *ecologically grown* were used to characterize organic production. Interpretation and application of these terms are often as difficult as defining organic. Some people say that organic farming is agriculture that is based on use of crop rotations, cover crops, composts, and nonchemical means of pest control and that excludes use of manufactured fertilizers or pesticides. That concept is limited in scope and does not cover the diverse practices and restrictions of organic farming.

In the early years of organic farming, about 1940 to 1970, the practice essentially consisted of local operations with much contact between consumers and farmers or retailers. With the expansion of organic production and markets in the 1970s, this association became more remote, and consumers, growers, and marketers needed means of proving that produce was organic. These needs led to the formation of certifying agencies that would vouch that the produce was grown organically and permit labeling of produce as being organically grown. By the 1980s, several private and state-run certifying agencies were operating in the United States. These agencies had variable standards for certification and charged variable fees for services. Some agencies did not recognize certifications by other agencies. Many regions of the country did not have locally accessible certifying agencies. These differences created problems of lack of certification, lack of uniform standards, unreasonable fees, and even fraud. To address these problems, the organic community sought federal legislation to enact national standards for certification of organic farming. This action led to the Organic Foods Production Act of 1990 and to the establishment of the National Organic Program (NOP) of the U.S. Department of Agriculture. The NOP sets regulations for certification. The NOP regulations are implemented through certifying agencies that are accredited by the NOP. Accreditation ensures that the certifying agencies understand and use the regulations of the NOP and conduct business properly. Certification applies to crops and crop products and to livestock, poultry, and products of their production. The certification practices referred throughout this book are those established by the NOP for the United States. Several hundred governmental and nongovernmental organizations in the world offer certification systems. Governmental organizations have increased in prominence and importance as the volume and value of organic production has increased.

The regulations of the NOP establish transition periods for movement from conventional farming into organic agriculture. A transition period is generally 3 years following the ending of applications of nonorganic practices. Some agencies may aid growers in selection of crops to grow during the transition period. In practice, growers may certify part of a farm and leave the rest in conventional agriculture. In that case, buffer zones between the organic farm and the conventional farm operations must exist. Regulations specify what physical distance or barriers are needed to separate the organic areas from conventional areas. Produce from mixed systems must be segregated at harvest. The regulations of the NOP also establish materials that are permitted, restricted, or prohibited for use in organic agriculture. Permitted materials can be used regularly in organic farming. Items that are restricted can be used only within the limitations set by the NOP standards. For example, farm manures need to be composted or an amount of time between application of the manure to cropland and harvest of produce must lapse to meet the organic standard. Prohibited materials cannot be used in organic production. Prohibited materials may be naturally occurring or manufactured. Consequently, definitions of organic agriculture by the NOP include listings of allowable practices in fertilization of crops, control of pests, and use of adjuvants (materials that affect the activity of other agents but that have little effect when supplied alone). Restriction of activities to those allowed in the listings and following specified practices of crop and soil management may permit a grower or the grower's produce to be certified as organic. Certification is done mainly by private organizations that work with development and review of allowable practices for growers who want to market their produce. Types of certification vary. Certification can be for products that are labeled as "100% organic," "organic," "made with organic ingredients," or "products with less than 70% organic ingredients." The grower or handler and certifier agree on a production or handling system that is appropriate for each classification. The certifier provides growers with definitions, guidelines, practices, and lists of materials that may be used in organic farming. A national list (NOP) of allowed and prohibited substances that can be used in organic farming is followed. This list includes natural and synthetic substances that are allowed or not allowed.

Growers that have sales of less than $5000 currently do not have to be certified by agencies to market produce as organically grown, but the production standards of organic farming should be followed.

ORGANIC MATERIALS REVIEW INSTITUTE (OMRI) PRODUCTS LIST

The *OMRI Products List* is a list of products that the Organic Materials Review Institute has determined are allowed for use in organic agriculture, including production, processing, and handling of farm-grown commodities. The products are reviewed for compliance with the policies of the USDA National Organic Program (NOP), and if they pass the review, the products are included on the OMRI Products List. Products on the list can carry the OMRI-listed seal, ensuring that the products listed meet NOP standards. The OMRI review program is private, and participation by providers is voluntary. Hence, a product may not be on the list and yet may meet NOP standards. Growers should consult the NOP list or their certifying agency for compliance of items not on OMRI Products List.

OTHER CONCEPTS AND BACKGROUND

Many people are not interested in certification for commercial organic agriculture and may want to develop their own working definitions. An understanding of *organic* as this term is defined by different scientists is helpful in the definition of *organic agriculture*. To biologists, *organic* means existing in or derived from a living organism. This definition is not unambiguous, for arguments may be made about the levels of existence in an organism, about contamination, and about modifications that may occur after the death of an organism. In chemistry, organic refers to the study of carbon-containing compounds with exception of some compounds, such as carbonates, which are considered to be inorganic compounds. Combination of these two definitions into one may indicate that organic farmers and gardeners can work with carbon-containing materials obtained from living organisms. Working with this combined definition, one could be a successful organic grower but not use the materials that would be permitted if the definitions were applied singularly. Limestone, largely calcium carbonate, is derived from living organisms and is an inorganic, carbon-containing compound. Diatomaceous earth for insect control is mostly silica and is strictly inorganic in chemistry, but is derived from unicellular algae, which are living organisms. Rocks and minerals, such as granite dust and greensand, are considered to be organic fertilizers by some people, and these materials were never living and are essentially void of carbon.

Organic farming has been defined as crop or animal husbandry with *natural materials*, whether these materials are from living or nonliving matter. The weakness of this definition in certification or in practice is in the use of the words *natural materials*. Not all natural materials are considered organic materials by the National Organic Program. On the one hand, potassium chloride is a naturally occurring fertilizer, but because of its high concentration of nutrients and high solubility, it does not qualify as being organic on the lists of certifying organizations. On the other hand, potassium sulfate and potassium magnesium sulfate (langbeinite) from the same mines as potassium chloride are considered to be organic materials. Cottonseed meal is suspected of contamination with insecticides used in cotton growing and is restricted in use in organic farming. Use of sewage sludge is prohibited.

Naturally occurring is a restrictive term and does not permit the use of manufactured materials, even though they may be identical to the materials produced in nature. For example, urea from the fertilizer plant is chemically equivalent to urea that is in urine. The latter source would be considered organic, whereas the former would not. Because of the potency of potassium chloride and fertilizer urea, these materials are not "organic" even though they are naturally occurring or identical to naturally occurring material. Some growers may recognize that the discrimination against these materials is due to the lack of understanding of how to use them. Growers who are educated on their use may feel that they can use potassium chloride and fertilizer urea and justify for their own interests that they are organic growers, but they cannot be certified as organic growers.

Strict definitions mandate that organic fertilizers be naturally occurring. A corollary to this definition is that the fertilizers be of low solubility or low in nutrients or both. Potassium chloride does not meet the stipulations of the corollary. Rock phosphate is naturally occurring and has a high phosphorus concentration but has low solubility and is therefore organic.

Physical treatment of materials normally is permitted. Rock phosphate is ground to silt-sized particles to increase the availability of its slowly soluble nutrients. Chemical treatment of rock phosphate with acids to manufacture superphosphates is not an organic practice, unless the treatment occurs in some natural process, such as mixing rock phosphate with decaying organic matter or mixing it in acid soils. Wood ashes are organic materials, although they are not naturally occurring in the context in which they are considered to be organic fertilizers. Hence, it is evident that a lot of judgment goes into the definition of organic fertilizers. The NOP publishes a national list of allowed and prohibited substances for use by certifying agents, and growers who are not seeking certification may want to consult this list.

Organic control of pests involves diverse activities. Generally, sprays or dusts derived from natural sources are considered to be organic. In some cases, a few manufactured, extracted, or purified products are permitted, as with oils and soaps used in insect control. Lack of mammalian toxicity or lack of injury to beneficial organisms is not necessarily a firm characteristic of organic pesticides. Organic sprays and dusts may be highly toxic, abrasive, or otherwise destructive to people, livestock, pets, fish, bees, and other organisms that are not targets for harm.

Organic control may involve cultural practices that limit the spread or growth of pests. These practices may include *biological control* or *natural control* of pests. Biological control involves the introduction of organisms that eat, kill, or impede the growth of pests. Natural control involves taking advantage of these organisms already existing in the environment without their intentional introduction. Use of barriers and traps, rotation of crops, drainage, fertilization, liming, and sanitation are cultural practices for pest control in organic agriculture. Many of these practices are in common with those used in conventional agriculture.

Many myths must be dispelled in discussions of organic agriculture. Organic agriculture is a highly managed system. An abandoned farm is not an organic farm. Failure to fertilize a crop is not an organic practice. Infestation of produce with insects is not a characteristic of organically grown material. Old or ancient practices are not necessarily organic ones. One cannot say that once upon a time, all farmers were organic farmers. Once upon a time, the most common practice was not to fertilize crops. Products such as lead or copper arsenate were used formerly for insect control. These products are not organic.

The organic farmer needs to be a scientist more than a philosopher. He must separate fact from myth and differentiate the occasional occurrence from the rule.

PRACTICES RELATED TO ORGANIC FARMING

Biodynamic farming and *permaculture* are systems of agriculture that have practices and concepts contributing to or added to foundations of organic farming. Biodynamic farming in a sense was a forerunner of organic agriculture as it was first outlined in the 1920s by an Austrian philosopher, Rudolph Steiner. The system of permaculture was developed by Australians Bill Mollison and David Holmgren in the 1970s. The word *permaculture* is derived from permanent agriculture and permanent culture.

Biodynamic Farming

In the 1920s, Rudolph Steiner outlined principles of biodynamic farming. A basic concept of biodynamic farming is the view of a farm as an organized, self-contained entity or as an organism with its own individuality. The farm is viewed as a closed, self-nourishing system. A biodynamic farm has an integration of crops with livestock and includes recycling of nutrients and management practices that involve environmental, social, and financial aspects of the farm. Although organic farming is basic to biodynamics, biodynamics is different in that it has an association with spiritual and astrological factors and emphasizes farming practices that have a balance between physical and higher, nonphysical realms that include the influence of cosmic and terrestrial forces on the farm. A certification system called *Demeter* was established in about 1924 for biodynamic farming.

Biodynamic farming can be divided into *biological* practices and *dynamic* practices. The biological practices are organic farming techniques that improve the quality of the soil. These practices include applications of farm manures and composts and growing of cover crops or green manures on the land. Biodynamic farming excludes the use of manufactured chemicals on soils and plants. Good tilth or physical structure of the soil is a factor that leads to production of high-quality crops thereby giving high-quality food for humans and high-quality feed for livestock. Biodynamic farming further involves adapting the farm to natural rhythms, such as planting crops during lunar phases, which are specific according to the type of crop—roots, fruits, or vegetation, for example.

The biodynamic system uses specific compost preparations for land application and other materials that are applied directly to crops. Biodynamic preparations are made of medicinal herbs that have already undergone a fermentation process to enrich them with growth-promoting substances. These preparations are added to active compost piles to speed and to direct the decomposition of plant materials in the pile and to preserve their original values. Other preparations are used as field sprays applied directly to the soil or crops. The benefit of the soil-applied sprays is to induce humus formation in the ground and to stimulate root growth. The crop-applied spray is used to improve the green color of plants and to prevent plant diseases. These sprays are said to harness cosmic forces that improve growth of plants. Development of the preparations involves specific practices, such as burying the components in animal parts, such as horns, bladders, or intestines, and leaving them in the ground over winter. Use of these preparations has been criticized as having no scientific basis, resembling alchemy, and as not contributing to the development of sustainable agricultural practices. In contrast with some applications in biodynamics, principles of organic farming generally have a scientific basis to support their use.

Permaculture

Permaculture is not a production system but is a land-use and community-planning philosophy. The term originally was derived from the term *permanent agriculture* but has evolved to stand for permanent culture. In the 1970s, Mollison and Holmgren developed ideas to create stable agricultural systems. They saw conventional, industrial agricultural methods as being harmful to soil, land, and water. Permaculture is based on creating ecological human settlements, particularly the development of a continual agricultural system that imitates the structure and interrelationships of natural ecosystems. The intent was that by training individuals in a core set of design principles, those individuals could design their own environments and build increasingly self-sufficient human settlements that would reduce reliance on industrial systems of production and distribution. Permaculture has a large international following of people who have received training through two-week-long courses in permaculture design.

2 Soil Fertility and Plant Nutrition

Soil fertility is a broad term that refers to the ability of a soil to supply nutrients to a crop. The ability to supply nutrients is affected by their amounts in the soil and is governed also by several other chemical, physical, and biological properties. Soil fertility is sometimes referred to as *soil quality* or *soil health*. These terms have been used to direct attention to the fact that soil fertility is related not only to the *chemical* factors of the soil but is governed by the *physical* and *biological* factors as well. *Chemical factors* of soil fertility refer to the supply of plant nutrients in the soil with regard to the total supply and to the amounts of the total supply that are available for plant nutrition. Soil acidity is a chemical factor of soil fertility. Soil acidity affects the solubility or availability of plant nutrients and other chemical elements that affect the ability of plants to grow in soils. Physical factors of soil fertility include conditions such as soil depth, water-holding capacity, drainage, aeration, tilth, temperature, and nutrient-holding capacity. Biological factors include conditions including the presence of harmful organisms such as plant diseases, weeds, and insects and the presence of beneficial organisms such as microorganisms that carry out mineralization of organic matter and nitrification of ammonium and that live in symbiosis with plants. Chemical, physical, and biological factors are interrelated, and sometimes it is difficult to place a soil property into one of the categories. Soil fertility is an integration of these factors.

Plant nutrition is the study of the uptake, transport, and function of nutrients in plants. Plant nutrients are known also as *essential elements*. These are chemical elements that are required for plant growth. They must be supplied whether the system is organic or otherwise. For a chemical element to be considered as a plant nutrient, several criteria must be met: (1) The element must be required for plants to complete their life cycles. Each element has a direct effect on plant growth or metabolism. Deficiency of an essential element will result in abnormal growth or premature death of plants. (2) The requirement for these elements is universal among plants; that is, all plants, not just a few, require these elements. If some plants require an element and others do not, the element is not considered to be a nutrient but a *beneficial element*. (3) No other element will substitute fully for an essential element. Partial substitution might occur among some elements, but each plant nutrient has a role for which no other element can substitute. Today, 17 chemical elements (Table 2.1) are recognized as essential for plants. Of these elements, three—carbon, hydrogen, and oxygen—are obtained from the air. Fourteen are derived from the soil. The requirement for each essential element is absolute, regardless of the source or amount required. Hence no nutrient is more important than another one, because each one is required absolutely for plant growth and metabolism.

Ninety-two naturally occurring elements are on the Periodic Table. Some people believe that each of these elements is essential for plant life and state that every natural element must be supplied for plant growth. This concept supports the recommendation that rock dusts are needed in plant nutrition to ensure adequacy of all of the naturally occurring elements. However, even if an element promotes plant growth but the criteria for essentiality listed above are not met, the element is not essential but is a *beneficial element*. Accumulation of an element in plants is not a criterion of essentiality or benefit since plants will absorb any element that is in solution and do not make absolute distinction between essential elements and any other element.

The list of essential elements is not static and may be expanded in the future as experimental techniques improve to purge elements from the environment in which plants are grown. Hydroponics was developed as a research tool in the mid-1800s to permit plants to be grown in media from which nutrients were added but from which an element under investigation was excluded. This procedure soon led to nitrogen, phosphorus, potassium, calcium, magnesium, sulfur, and iron being added to the list of essential elements. Carbon, hydrogen, and oxygen were identified earlier, before the development of hydroponics, as essential, with these elements being identified as constituents of water or organic matter. In the 1920s, techniques for research on plant nutrients improved, and many other elements were added to the list. Molybdenum was added in 1939, and chlorine was added in 1954. The last element to be accepted as an essential element is nickel in 1987. Perhaps additional elements, including some of the beneficial elements, will be accepted as essential in the future.

One nutrient is not more essential than another, but the elements are not required in the same amounts by plants. Because of differences in amounts of requirements, elements are divided into classes of *macronutrients* and *micronutrients* (Table 2.1). On a dry weight basis, the concentrations of macronutrients in plants range from about 0.3% to 5% or higher, depending on the plant part and species under consideration.

The concentrations of nutrients listed (Table 2.1) are only guidelines, and the concentrations of nutrients in plants can vary widely from those listed depending of the plant part, plant age, soil fertility, and many other factors. Nitrogen is normally the most abundant of the soil-derived elements in plants, followed by potassium (K), calcium (Ca), magnesium (Mg), phosphorus (P), and sulfur (S) in that general but not absolute order. Micronutrients on an analytical basis are minor constituents, ranging from one to several hundred parts per million of the plant dry weight. For comparison of concentrations between macronutrients and micronutrients, 1% is 10,000 parts per million. The micronutrients are vital to plant growth and development and are just as important as the macronutrients in plant health. It is redundant to refer to plant nutrients as essential nutrients, as all nutrients are essential by definition.

Some *beneficial elements* stimulate plant growth. These elements do not meet all of the requirements of essentiality—that is, they have not been shown to be required by all plants; distinct metabolic roles have not been demonstrated; or the requirement is not absolute. Sodium, silicon, selenium, cobalt, and, perhaps, aluminum, vanadium, and others, are beneficial elements.

TABLE 2.1
Listing of Essential Elements (Plant Nutrients) and Their Approximate Concentrations (Dry Weight Basis) in Plant Foliage

Essential Element					
Obtained from Air		Obtained from Soil			
Nutrient	(%)	Macronutrients	(%)	Micronutrients	(ppm)
Carbon (C)	50	Nitrogen (N)	3.0	Iron (Fe)	100
Hydrogen (H)	5	Phosphorus (P)	0.4	Zinc (Zn)	20
Oxygen (O)	40	Potassium (K)	2.0	Copper (Cu)	5
		Calcium (Ca)	1.0	Manganese (Mn)	50
		Magnesium (Mg)	0.5	Molybdenum (Mo)	0.1
		Sulfur (S)	0.3	Boron (B)	30
				Chlorine (Cl)	100
				Nickel (Ni)	1

FERTILIZERS

Most soils do not have an unlimited capability to supply essential elements to crops. *Fertilizers* are materials that are used to carry plant nutrients to soils. About 70% of the time, increases in crop yields are recorded after soils are fertilized with nitrogen. About 40% to 50% of the time, increased yields occur after fertilization with phosphorus and potassium. These frequencies of increased yield responses are high enough that most of the time, growers add nitrogen, phosphorus, and potassium fertilizers to their land. Growers use fertilizers containing calcium, magnesium, or sulfur to enrich soils in areas where these elements are deficient, but these deficiencies do not occur with the frequencies as those of nitrogen, phosphorus, and potassium. Deficiencies of micronutrients are much rarer than those of macronutrients, and specific applications of fertilizers for micronutrients are not common practices. The beneficial elements, such as silicon, cobalt, sodium, and selenium, are not applied to crops or soils or tested for in most situations of farming and gardening. The beneficial elements are generally ubiquitous in the environment.

Sandy, acid, or leached soils are likely to be deficient in all essential elements. In these cases, the nutrients were in low amounts in the soils since the beginning of the soil-forming processes or have been depleted from soils by natural forces, such as leaching with water. Minor elements may be deficient in alkaline soils ($pH > 7.5$) and in organic soils (peats, mucks) because the elements are combined with soil materials and are not soluble or are not in an available form for acquisition by plants. The amounts of nutrients held in soils vary with factors such as texture (proportions of sand, silt, and clay) and with the amounts of organic matter in soils. Biological activity is required to release nutrients from organic matter by a process known as *mineralization*. The release of nitrogen from organic matter is referred to as *nitrogen mineralization* or *ammonification*. Ammonia is the first product formed from nitrogen mineralization of organic matter. Soon after release ammonia is oxidized by microorganisms to nitrate, and this process is called *nitrification.*

Organic Fertilizers

Use of the term *organic fertilizers* is subject to their definition. A working definition of *organic fertilizers* is needed for convenience in discussing these fertilizers. Organic fertilizers are naturally occurring materials of biological or mineral origin and are low in nutrient concentrations or solubility or have both properties. Organic fertilizers may be altered physically in processing for agricultural uses, but chemical processing usually does not occur. *Chemical* fertilizers, *synthetic* fertilizers, and *conventional* fertilizers are synonymous terms. Conventional fertilizers may be manufactured from raw materials having no initial value as fertilizers (nitrogen from the air, for example), or they may be manufactured by purification or chemical treatment of naturally occurring materials that in their native state are organic fertilizers (rock phosphate treated with acids, for example, to make superphosphates). Purification and chemical processing usually increase the concentration or solubility of nutrients in fertilizers.

Although this fact has not always been recognized, plants absorb soil-borne nutrients only when the nutrients are in aqueous solution. In the 1700s, it was believed that the purpose of tillage was to break the soil into small particles so that mouths of the roots could eat and gain nourishment from the soil. Also, nutrients are not taken up by direct exchange between roots and solid particles in the soil, although this concept has been difficult to dispel. With the probable exception of boron, nutrients in aqueous solution are in ionic form, that is, they are charged particles in solution (Table 2.2). These ions are the forms of nutrients that are immediately available to plants.

For a plant to get nutrition from a fertilizer, the fertilizer must be dissolved so that its essential elements are in one of the forms in Table 2.2 or in a state that can be converted easily to these forms. Nutrients in organic fertilizers are converted slowly to soluble forms, whereas chemical or conventional fertilizers may dissolve rapidly in water. Organic fertilizers require biological action or strictly physical or chemical

TABLE 2.2
Ionic Forms of Nutrients That Plants Absorb from the Soil Solution

Nitrogen: NO_3^- (nitrate); NH_4^+ (ammonium)	Iron: Fe^{2+} (ferrous ion)
Phosphorus: $H_2PO_4^-$ (monobasicphosphate)	Zinc: Zn^{2+} (zinc ion)
Potassium: K^+ (potassium ion)	Copper: Cu^{2+} (cupric ion)
Calcium: Ca^{2+} (calcium ion)	Manganese: Mn^{2+} (manganous ion)
Magnesium: Mg^{2+} (magnesium ion)	Molybdenum: MoO_4^{2-} (molybdate)
Sulfur: SO_4^{2-} (sulfate)	Chlorine: Cl^- (chloride)
	Boron: H_3BO_3 (boric acid); BO_3^{3-} (borate)
	Nickel: Ni^{2+} (nickel ion)

action for their elements to be released to ionic forms. Fertilizers composed of organic matter must be decomposed by microorganisms to form ions. This biological decomposition is called *mineralization*. Rocks and minerals are decomposed by physical and chemical actions that are termed *weathering*. The following reactions illustrate the processes by which organic fertilizers become soluble or available to plants.

Mineralization: Biological action

$$\text{Organic Matter} \rightarrow NH_4^+;\ NO_3^-;\ H_2PO_4^-;\ SO_4^{2-};\ Ca^{2+}$$

Weathering: Physical and chemical processes

$$\text{Rocks, Minerals} \rightarrow \text{Small particles} \rightarrow K^+;\ Ca^{2+};\ Mg^{2+};\ H_2PO_4^-;\ SO_4^{2-}$$

A common but not universal trait of chemical fertilizers is their rapid availability or high solubility in water.

Chemical fertilizers: Dissolution in water

$$\text{For example, ammonium nitrate, } (NH_4NO_3) \rightarrow NH_4^+ + NO_3^-$$

$$\text{Concentrated superphosphate, } [Ca(H_2PO_4)_2] \rightarrow Ca^{2+} + 2H_2PO_4^-$$

$$\text{Potassium chloride, } (KCl) \rightarrow K^+ + Cl^-$$

As mentioned previously, plants obtain nutrition from dissolved substances in the soil solution. The same kinds of ions are formed whether the initial material was an organic or chemical source, but the rate of formation of soluble substances differs. Once these ions are formed, plants do not distinguish whether the ions were of organic or chemical origin. The differences in organic and chemical fertilizers in supplying nutrients to crops are not in the kinds of nutrients supplied but in the rates of production of the available or soluble nutrients. Generally, organic fertilizers release nutrients slowly and in response to environmental factors, such as soil moisture and temperature, and in response to the effects of these factors on microbial activity in the soil. Chemical fertilizers, however, release nutrients rapidly with less dependence on environmental factors other than water supply and perhaps temperature.

Sometimes chemical fertilizers are manufactured to mimic the slow release of organic fertilizers. These fertilizers are called *controlled-release* fertilizers or *slow-release* fertilizers. The properties of slow release arise from coating the fertilizers with plastic or sulfur or from embedding the fertilizers in a sparingly soluble matrix that slows the dissolution of the nutrients and resists the effects of microorganisms on the fertilizers. Sometimes chemical agents are included with the formulation of fertilizers to slow or inhibit microbial action. These treatments of chemical fertilizers substantially increase their prices.

The advantages and disadvantages of organic fertilizers in relation to chemical fertilizers are based on the release of nutrients from individual materials. The following statements, with qualifications that depend on the type of organic material, list some advantages and disadvantages of organic fertilizers.

Advantages of Organic Fertilizers

1. Organic fertilizers commonly are mild, noncaustic materials, and if they come in contact with a crop, the fertilizers likely will not burn or desiccate foliage or roots.
2. The slow release of nutrients makes them available for a longer period of time than water-soluble chemical fertilizers, which may be leached with the downward movement of water.
3. Use of organic fertilizers with high organic matter contents can improve the physical properties of soils in ways such as imparting higher water-holding capacity and better structure and good tilth or physical condition of soil for crop growth. Some of the organic matter of fertilizers is converted to humus in soil by a process known as *humification*.
4. Organic materials are sources of many essential elements.
5. The use of organic fertilizers, such as composts, is a method of recycling materials that might otherwise be wasted.

Disadvantages of Organic Fertilizers

1. Organic fertilizers that contain low concentrations of nutrients must be obtained and applied in large quantities to deliver sufficient nutrients to grow a crop. The materials may be bulky and difficult to apply to the soil.
2. Because of their slow release of nutrients, some organic fertilizers may not supply nutrients sufficiently rapidly or in large enough amounts to support the demands of a crop. The release rate may be too slow to provide nutrients to a crop that has been diagnosed as being nutrient-deficient. Poor solubility of the fertilizers may limit dispersion of the nutrients in the soil.
3. Although they may contain a multitude of elements, the concentrations of some nutrients in organic fertilizers may be too low to be of value to a crop. The supply of nutrients may not be balanced to meet the needs of crops. Good sources of organic fertilizers for P and K are scarce.
4. Even if they are raised on the farm, organic fertilizers usually are more expensive than chemical fertilizers, with the exception that some of the slow-release chemical fertilizers may cost as much as organic fertilizers.

Chemical fertilizers were developed in response to the apparent disadvantages of organic fertilizers. Because of their slow-release properties, organic fertilizers may not supply sufficient nutrition rapidly enough to a crop to prevent or alleviate deficiencies. Chemical fertilizers can provide an immediately available source of nutrients, and plant response to these fertilizers is limited only by the capacity of the plants to absorb and assimilate the nutrients. Because of their generally low concentration of nutrients, the bulk of organic fertilizers makes them inconvenient to handle. An 80-pound bag, or less, of chemical fertilizer can contain as much or more nutrients than a ton of farm manure or compost. Commercial chemical fertilizers are readily available in the market. Organic fertilizers often are more difficult to obtain than chemical fertilizers and usually bear a high price. Some organic fertilizers, dried blood, for example, are very expensive because of their scarcity. Also, much of the cost of fertilizers is in their shipping and handling. Shipping costs for a low-analysis material will be as much as for a concentrated material.

Growers may find it practical to use a balanced approach in uses of organic and chemical fertilizers. Organic fertilizers may be used for building soil fertility over a long period of time. Initially, growers may rely on supplemental use of chemical fertilizers with organic fertilizers in poor soils. As the fertility with respect to nutrient supply increases with prolonged application of organic materials, use of chemical fertilizers may be tapered off. A grower may use chemical fertilizers in the making of organic fertilizers. Chemical nitrogen fertilizers often are added to compost piles to accelerate the rate of decomposition of organic matter. Chemical phosphate fertilizers also are added to composts and to farm manures to fortify the phosphorus concentrations of these otherwise phosphorus-poor materials. Certified organic growers, however, have limitations on the use of chemical fertilizers in crop production. Certifying organizations require a time period of 3 years between the discontinued use of chemical fertilizers and issuance of certification as an organic grower.

ANALYSES OF FERTILIZERS

Fertilizers that are sold commercially, whether organic or chemical, are labeled for their nutrient contents. The *analysis* of a fertilizer is a term that is synonymous with *grade* of fertilizer. The analysis or grade is expressed by three numbers that indicate the nitrogen, phosphorus, and potassium in the fertilizer. The label on the fertilizer container reports the analysis as *available nitrogen* (N), *available phosphoric acid* (P_2O_5), and *available potash* (K_2O). Note that only the nitrogen is reported on an elemental basis. Soil tests and recommendations for fertilization of crops are coordinated with these chemical expressions of nutrient content. The term *available* is vague, but in the case of fertilizers, practically, it means the amount of nutrient in the fertilizer that a crop could obtain from the fertilizer in one growing season. Solubility and mineralization of fertilizers, as well as concentrations, are among characteristics that govern nutrient availabilities.

Expression of the analysis as oxides of phosphorus and potassium is confusing in that some people equate these values with actual phosphorus and potassium concentrations. This view is not a serious problem, for as previously mentioned, recommendations for fertilization are based on the grade of fertilizer as expressed as oxides of phosphorus and potassium. No correction is needed to express nitrogen as actual nitrogen, for analyses of fertilizers express nitrogen as the element N. A correction of 0.436 times the available phosphoric acid concentration is needed to express the actual phosphorus in a fertilizer. A correction of 0.830 times the available potash is used to express actual potassium in a fertilizer. A fertilizer with 10% N - 10% P_2O_5 - 10% K_2O converted to its actual nitrogen, phosphorus, and potassium concentrations is 10.0% N - 4.36% P - 8.3% K by weight.

TABLE 2.3
Comparison of Retail Costs of Nutrients in Commercial Dried Cow Manure (2-1-2) and a Mixed Grade (10-10-10) Chemical Fertilizer

	Fertilizer	
Costs	Cow Manure (2-1-2)	Mixed Grade (10-10-10)
Per hundred wt	$12.00	$20.00
Per lb N	6.00	2.00
Per lb P_2O_5	12.00	2.00
Per lb K_2O	6.00	2.00
Per lb total nutrients	2.40	0.67

Costs of fertilizers can be calculated based on their price per unit or pound of N, P_2O_5, or K_2O or on the total nutrient supply. A comparison of two commercial fertilizers, one organic and one chemical, is made in Table 2.3. Prices are for comparisons only and will vary with location of purchase and with the amounts of fertilizers purchased.

In considering the costs of fertilizers, the grower must evaluate the purpose for which the fertilizer is used against the cost of nutrients. For example, in the case with the comparisons between cow manure and dried blood (Table 2.4), the cheaper price of cow manure may not be sufficient justification to opt for this fertilizer instead of dried blood. The release of nitrogen from cow manure is much slower than that from dried blood. Less than half of the nitrogen from cow manure will be available to a crop in the first growing season, whereas nearly all of the nitrogen in dried blood will be released. Also, the slow release of nitrogen from cow manure limits its use to correct nitrogen deficiency in a growing crop, but the dried blood provides a sufficiently quick release of nutrients to restore nutrition to a nitrogen-deficient crop. The

TABLE 2.4
Comparison of Costs of Nutrients in Commercial Dried Cow Manure (2-1-2), Seed Meals (6-3-0.5), and Dried Blood (12-0-0)

	Fertilizer Cost		
Unit	Dried Cow Manure (2-1-2)	Seed Meals (6-3-0.5)	Dried Blood (12-0-0)
Per hundred wt	$12.00	$18.00	$80.00
Per lb N	6.00	3.00	6.67
Per lb total nutrients	2.40*	1.89*	6.67
Mineralization	Slow (<50%**)	Fast (60–80%**)	Fast (90%**)

* Includes available phosphoric acid and potash as shown in Table 2.3.
** Estimated mineralization rate for nitrogen release in one growing season.

quick release (mineralization) of nitrogen from dried blood may create a problem with ammonium accumulation in soils in which seeds or roots of seedlings may be in close contact with the fertilizer. A waiting period of 1 or 2 weeks between fertilizing and planting is recommended to avoid the potential of injury from the ammonium from rapidly mineralizing fertilizers. In comparison to cow manure and dried blood, about 60% to 80% of the nitrogen in seed meals (cottonseed meal, linseed meal, soybean meal, castor pomace) would be available. A release rate of 60% to 80% is considered rapid. Release of nutrients from composts is slow.

Caution must be taken when adding organic matter that has a high ratio of carbon to nitrogen. These organic materials are called *carbonaceous materials* and are not fertilizers. Strawy manure, unfinished compost, sawdust, wood chips, bark, and paper are examples of carbonaceous materials. If these materials are added to soil, they do not contain enough nitrogen for their mineralization or decomposition. The microorganisms that are doing the decomposition will use the available soil nitrogen in the process. This microbial consumption of nitrogen or other nutrients is referred to as *immobilization*. The immobilized nutrients are not available for plant nutrition until the microorganisms die and are in turn mineralized.

3 Requirements of Plants for Soil-Derived Nutrients

This chapter presents information on (1) the functions of nutrients in plants, (2) the effects of nutrients on plant growth and quality, (3) recognition of symptoms of deficiencies of nutrients, and (4) how to supply nutrients to plants. For most of the nutrients, specific metabolic functions in plants have been identified. Participation in these metabolic roles is a factor that makes an element essential. Because of the metabolic disorders associated with shortages of a nutrient, limitations in supply of any nutrient may restrict plant growth, development, and yields and cause appearance of symptoms of deficiency. Often deficiencies of nutrients are expressed in lower quality of produce. Increasing the supply of the nutrient will enhance growth and yields within limits and also will have effects on crop quality, for example, developing green color in a leafy vegetable crop. However, supply of nutrients in excess of the needs of a crop may have an adverse effect on crop quality, often lowering quality or suppressing harvest yields.

Severe shortages of nutrients usually lead to development of symptoms of deficiency. Recognition of these symptoms is a useful way of identifying nutritional disorders in a crop. If the deficiency is detected in time, fertilization may restore crop productivity. If the deficiency is recognized too late for correction in the current crop, the grower is alerted that remedies need to be taken for the next season.

Fertilizers are materials that carry plant nutrients to the soil. This chapter will present and evaluate organic and chemical fertilizers for each of the plant nutrients and will discuss practices that increase the nutrient-supplying capacity of soil.

NITROGEN

FUNCTIONS

The discovery of the essentiality of nitrogen is attributed to Theodore de Saussure, who in 1804 published his research that showed that normal growth of plants was not possible without the absorption of nitrates and other minerals from the soil. Nitrogen has many functions in plants. It is a component of proteins, genetic material (the nucleic acids DNA and RNA), chlorophyll, and many other compounds that are vital in plant metabolism. Proteins are nitrogen-rich compounds and are major nitrogenous constituents of plants. By weight, about one part in six of the average protein is nitrogen. About 85% or more of the nitrogen in plants is in protein. Another 10% of the total nitrogen in plants is in soluble nitrogenous compounds, such as uncombined amino acids and unassimilated nitrate and ammonium. The remaining 5% or less of the total nitrogen is in the genetic material, chlorophyll, coenzymes of metabolism (vitamins and the like), and lipids, among other compounds.

Effects of Nitrogen on Plant Growth and Quality

Nitrogen is a potent nutrient, the deficiency of which can severely limit crop production. Application of nitrogen fertilizers must be monitored closely to avoid underfertilization or overfertilization. Nitrogen is deficient in about 70% of crop land. Recovery of growth and yield potential from nitrogen deficiency during resupply of the nutrient can be rapid; however, over-application of nitrogen may have adverse effects on growth and quality. Some of the effects of limited, optimum, and excessive nitrogen fertilization follow.

Vegetative Growth

Applications of nitrogen fertilizers to plants promote vegetative growth (leaves, stems, roots) and reproductive growth (flowers, fruits, seeds), but stimulation of vegetative growth generally exceeds that of reproductive growth Also, shoot growth is enhanced more than root growth. Suppressions of root growth and reproductive growth by excessive fertilization occur at lower applications of nitrogen than with stem and leaf growth. These responses are due to the effects that nitrogen has on the hormonal balance of plants. Nitrogen promotes the biosynthesis of growth-regulating compounds that stimulate shoot growth and that inhibit root and reproductive growth.

With crops that are grown for their vegetation, such as spinach, lettuce, and celery, the promotion of vegetative growth by nitrogen is a favorable response. However, with root crops, for example, beets and carrots, stimulation of shoot growth may be so strong that yields of roots are diminished by application of nitrogen. With fruit-types of vegetables, tomatoes, for example, stimulation of growth of stems and leaves by overly generous applications of nitrogen fertilizer may dominate and persist for so long that yields of fruits are reduced relative to those obtained with optimum fertilization. No advantage is gained by pruning of the vegetative growth to try to bring

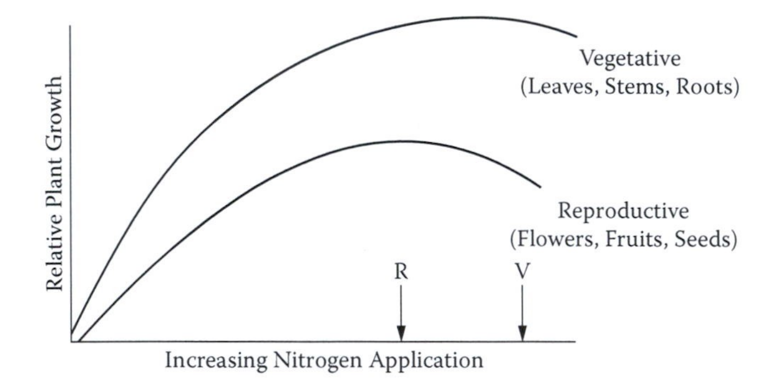

FIGURE 3.1 Relative growth of plants in response to nitrogen fertilization. Upright lines indicate relative levels of fertilization associated with depression of reproductive growth (R) and vegetative growth (V).

production of vegetation in balance with flowering and fruiting. Vegetative growth is still likely to dominate after pruning so that the first new growth after pruning is vegetative. Pruning in this case weakens the plant and further retards reproductive growth.

Effects on Succulence

Nitrogen fertilization increases the proportion of water in plants and increases the succulence or juiciness of vegetative parts. Succulence is a desirable feature in many vegetables—lettuce, spinach, radish, celery, and others for which the vegetative portion is edible portion. Possibly, the succulence of fruits can be raised by nitrogen fertilization, but the effects will be smaller than those on the vegetation.

Effects on Cell Walls

Cell walls of plants that are well nourished with nitrogen are thinner than those receiving lesser amounts of nitrogen. Nitrogen in plants promotes protein synthesis at the expense of carbohydrate synthesis and accumulation. Cell walls are made of carbohydrates or of materials derived from carbohydrates. The diversion of carbohydrate to protein lessens the amount of material available for synthesis of constituents (cellulose, lignin, for example) of cell walls; consequently, cell walls do not thicken as much with an abundant supply of nitrogen as they would under situations of lesser abundance of nitrogen (Figure 3.2). In addition, protein synthesis increases the protoplasmic portion of cells and increases the amounts of water in cells. The pressure caused by the force (called *turgor pressure*) of the increased water on the cell walls in interaction with other growth factors causes the walls to stretch and to become thinner.

The thin cell walls and high water content give characteristics of succulence and crispness to vegetables. These vegetables are not fibrous or tough. All vegetative portions of plants may be affected in this way. The stretched-out, thin-walled cells of stems are weak. Consequently, the stems are weak and may not be able to support plants in an upright position.

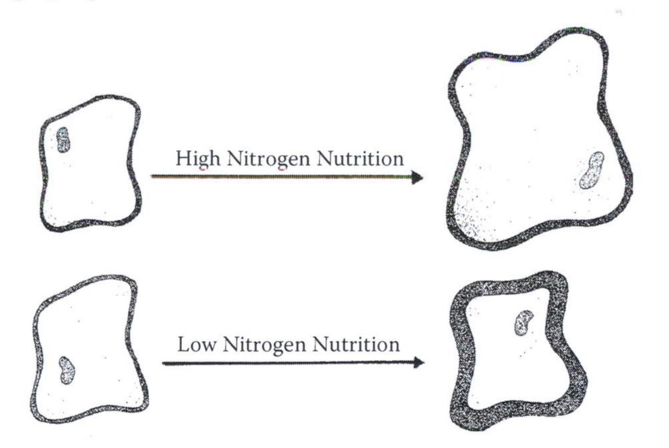

FIGURE 3.2 Plant cells growing with a high level or with a low level of nitrogen nutrition. Cell walls of plants receiving high amounts of nitrogen are thinner than those receiving low amounts of nitrogen fertilization.

The displacement of a plant from its upright position is called *lodging* (Figure 3.3). Lodging of plants can cause losses of produce by the material falling on the ground and becoming dirty, rotting, or being below the level of harvesting equipment. Lodging is a particular problem with small grains—the cereals, wheat, oats, barley, and rye. Fairly recent developments of dwarf wheat and rice varieties that are stiff-stemmed and that respond without lodging to fertilization with nitrogen have allowed for raising of productivity of cereal grains. The development of these grains that were responsive to high amounts of nitrogen fertilization was a part of the Green Revolution, introducing an intensive form of agriculture that raised world-wide productivity of crops.

Plant organs that contain these thin-walled cells in addition to being structurally weak are bruised easily. Considerable damage by breakage of leaves and stems may occur during harvest. The high water and protein contents and thin-walled nature of cells makes plant organs susceptible to attack by insects and diseases. These cells offer little resistance to attack and offer excellent nutrition for insects and diseases.

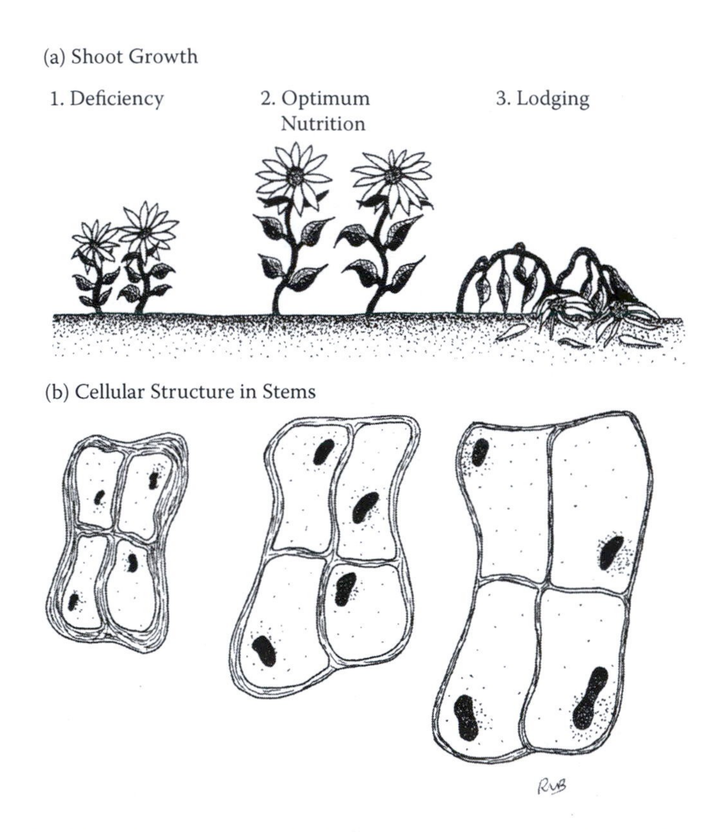

FIGURE 3.3 Plant shoot growth and cellular structure in stems of (1) unfertilized plants, (2) optimally fertilized plants, and (3) overfertilized plants (with lodging). Note that cell walls progressively become thinner as nitrogen fertilization increases.

Effects on Plant Maturation

The maturation of a crop is slowed by nitrogen fertilization. Nitrogen fertilization must be appropriate for the kind of crop that is being grown, that is, whether the crop is grown for its vegetation or fruits or seeds and whether the crop has a short or long season for maturation. Through its effect on growth regulators (hormones), nitrogen fertilization enhances and prolongs the vegetative stage of plant development. Flowering and fruiting may be delayed by nitrogen fertilization.

Ample nitrogen fertilization is needed for development of a good vegetative frame. Plants that are undernourished will be weak and spindly and will be unable to support their flowers and fruit. Undernourished plants will have a short period of productivity, and yields will be curtailed by senescence or aging of the plants prematurely. Nitrogen fertilization will permit development of strong plants that can support a high level of flowering and fruiting and maintain a long period of productivity. With varieties of tomato and cucumber that flower and fruit over a period of time, nitrogen fertilization will prolong this period of productivity so that yields are enhanced above those of underfertilized crops. However, overabundance of nitrogen may delay reproductive growth so that it occurs so late in the season that little productivity is obtained before cold weather or frost. In extreme cases, reproductive growth is delayed so long that none occurs before frost or that which occurs is insufficiently mature to give a product that can be harvested.

Symptoms of Nitrogen Deficiency in Crops

Nitrogen is essential for syntheses of proteins and chlorophyll. With shortages of nitrogen, syntheses of these compounds will be restricted. These restrictions in plant metabolism will appear in the development of symptoms of nitrogen deficiency (Figure 3.4). Nitrogen-deficient plants are *spindly*, meaning slow-growing, weak-growing, or stunted, and off color. If nitrogen was limited from the initiation of growth soon past the seedling stage, the entire plant will be light-green or yellow-green. If the supply of nitrogen is exhausted after considerable growth is made, the lower leaves will have a light-green or yellow-green coloration. Eventually, the lower leaves will turn brown and drop off, unless the supply of nitrogen is restored.

Nitrogen is a *mobile element* in plants, meaning that it can be transported readily from one part of a plant to another part. Nitrogen can be mobilized or transported from the old leaves of the plant to the young growing regions or reproductive organs of plants. The resulting transport of nitrogen from the old leaves to the growing regions and reproductive organs results in depletion of nitrogen and in the appearance of nitrogen deficiency symptoms in the old leaves. The yellowing and stunting of plants are identifying characteristics of nitrogen deficiency. Often, during nitrogen deficiency, lower leaves may become reddened, particularly along the veins. Sometimes, this symptom is confused with that of phosphorus deficiency, which also produces reddening of lower leaves, and experience is needed to learn how to differentiate nitrogen deficiency from phosphorus deficiency. The nitrogen-deficient foliage will have an overall lighter color than the phosphorus-deficient foliage. Reddening of phosphorus-deficient foliage will be spread across the leaf blades and

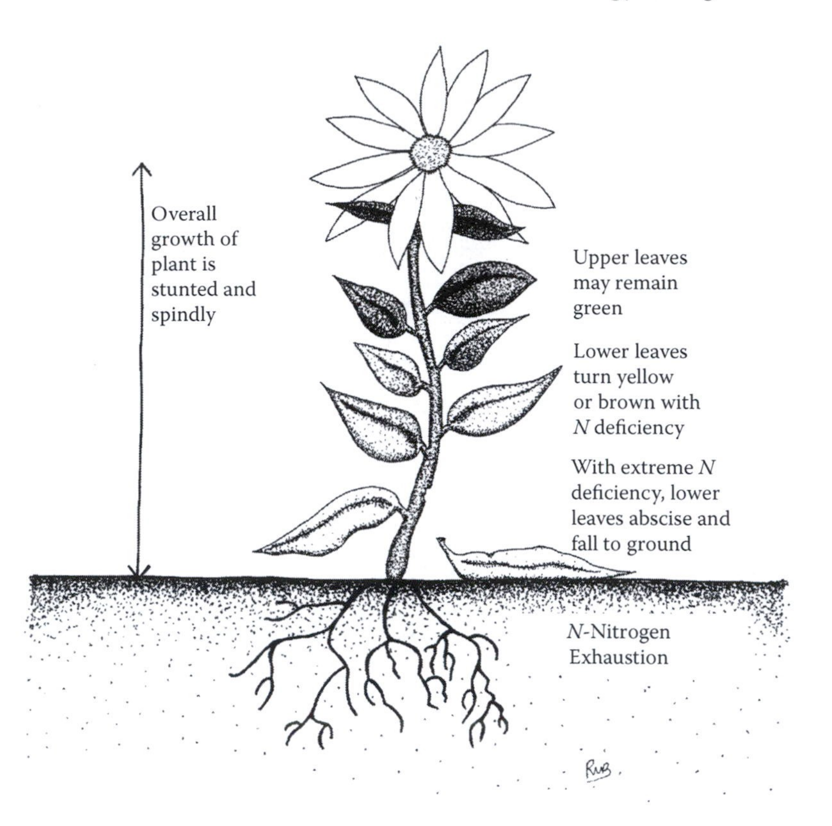

FIGURE 3.4 Expression of nitrogen deficiency symptoms on shoots of plants growing in soil in which nitrogen is depleted during the growth of plants.

will be strongly evident on the underside of the leaf, whereas the reddening during nitrogen deficiency will be concentrated along the veins and will be equally noticeable on the top or bottom of the leaves.

Nitrogen deficiency during vegetative growth will force plants to mature earlier with significant losses of quality and yields. Number and size of flowers and fruits will be smaller in a nitrogen-deficient crop than in a nitrogen-sufficient crop. Root growth will be restricted greatly by nitrogen deficiency, for roots will not grow in soil zones that are nitrogen deficient. Although nitrogen has differential effects on growth of different plant parts, it is needed for growth of all plant parts.

Readily available nitrogen applied to the soil can be absorbed rapidly by a plant. Deficiencies that appear during the early stages of vegetative growth may be corrected with only a small loss of yield potential. Deficiencies that appear in late stages of vegetative growth can be corrected, but yields will be limited substantially by the deficiencies. Deficiencies that appear near or at the time of flowering or fruiting are not likely to be corrected during the current season and can limit yields substantially.

Often as plants mature, nitrogen is transported from the vegetative organs to the reproductive organs. This process is a natural one in which nitrogen accumulated during vegetative growth is used later in reproductive growth. Therefore, appearance of symptoms of nitrogen deficiency on foliage during crop maturation is not always an indication that nitrogen supply in the soil was insufficient.

Amounts of Nitrogen Required by Crops

Crops differ in the amounts of nitrogen that they remove from the soil in plant growth and development. The amount of nitrogen removed is a function of the amount of dry matter that a crop produces and of the nitrogen concentration in the dry matter. Well-nourished plants have from 1% to 4% nitrogen or more in their foliage, with variation occurring among species and with plant age. Nitrogen-deficient plants may have less than 1% nitrogen in their foliage. High-yielding crops—those producing large amounts of vegetative growth, fruits, or seeds—have heavy requirements for nitrogen. A fast rate of growth in conjunction with a large amount of growth demands a rapidly and abundantly available supply of soil nitrogen. Crops that are grown at high densities remove large amounts of nitrogen from the soil because of the high productivity of crops grown in dense plantings. High-yielding crops that have high protein contents have high demands for soil nitrogen. Some examples of crops with high demands for nitrogen from the soil are as follows.

High Demand for Nitrogen

More than 120 lb N per acre (3 lb N per 1000 sq ft) removed in one season.

Corn
Potatoes
Tomatoes
Celery
Forages (hay, pasture)

Crops with moderate demands for nitrogen produce less dry matter than those with the high demands for nitrogen. These plants may be inherently low yielding or may be grown at lower densities of planting than those with high demands for nitrogen. Most vegetable crops fall into this category. Some crops that have moderate demands for soil nitrogen are listed as follows.

Moderate Demand for Nitrogen

Between 50 and 120 lb N per acre (1.25 to 3 lb N per 1000 sq ft) removed in one season. Some crops with moderate nitrogen requirements include

Most vegetable crops
Short-stemmed cereals (wheat, oats, barley, rye)
Garden legumes (peas, beans)
Turfgrasses

Crops with low demands for nitrogen from the soil include those that are slow growing and those that have a very short growing season. Many root crops have low requirements for nitrogen. Overfertilization of root crops with nitrogen may promote shoot growth and inhibit root growth with the result that root yields are considerably lower than those produced with optimum fertilization. Crops that are grown for their sugar contents remove low amounts of nitrogen in proportion to their yields. Generous nitrogen fertilization of these crops will enhance dry matter production but will lower the sugar yields because of the diversion of carbohydrate to protein. Some crops that remove small amounts of nitrogen from the soil follow.

Low Demand for Nitrogen

Less than 50 lb N per acre (1.25 lb N per 1000 sq ft) removed in one season.

Tobacco
Root crops (carrots, turnips, sweet potatoes)
Radish
Tall-stemmed cereals
Orchards

The amounts of fertilizer applied to crop land also vary with soil types, weather, production goals, management practices of growers, varieties of crops, and other factors. Applications of nitrogen from fertilizer may be 50% to 100% greater than the amount that is removed by a crop to compensate for losses and failures of plants to recover nitrogen that is applied. Some of the applied nitrogen may be lost by leaching into the ground, volatilization into the air, denitrification (biological conversion of nitrate to gaseous forms of nitrogen), competition from weeds and microorganisms; some of the nitrogen may be placed out of reach of plant roots, as in between rows of widely spaced plantings. Plants typically recover 50% to 70% of the nitrogen applied in fertilizers, although the recovery may be much less than 50% if applications are high, weeds are competitive with the crops, and conditions for loss are favorable. Losses of nitrogen can be limited by conservation practices, such as supplying nitrogen in several applications, called *split applications*, during the season and placement of fertilizer in bands along rows of crops rather than broadcasting it between rows (Figure 3.5).

Split applications of nitrogen help to supply nitrogen to crops as the need for nitrogen arises. One-time applications at the beginning of the season are subjected to losses by leaching and other environmental factors. Costs of multiple applications are deterrents to use of this practice relative to applying all of the fertilizer at planting. Placing the fertilizers in bands locates the nutrients in close proximity to the roots and limits the losses to leaching and to weeds. Banded nitrogen may be applied all at planting or during the growing season by sidedressing.

Growers using organic fertilizers that release nitrogen slowly should apply nitrogen in amounts that are related inversely to the rate of release of nitrogen; that is, fertilizers that release nitrogen slowly usually are applied in large quantities. Nitrogen from these slow-release fertilizers may carry over to the next growing season, and a downward adjustment, based on experience or expert advice, can be made in the second and subsequent years of fertilization. The buildup of available nitrogen from slow-release organic fertilizers is slow, and therefore many people tend to overvalue the benefits of the residual effects of fertilizers. An indication of the residual value

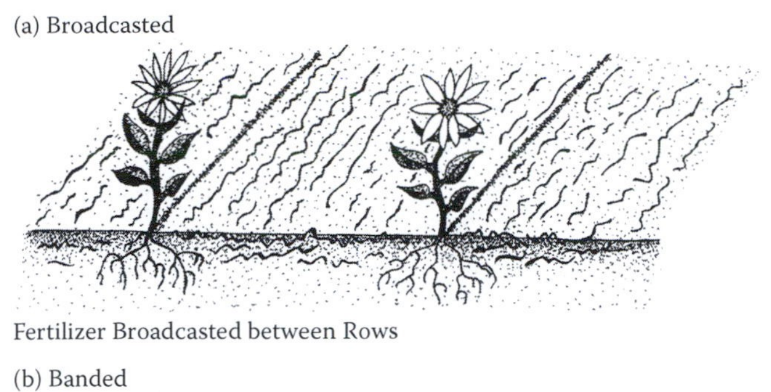

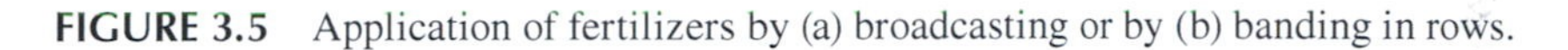

FIGURE 3.5 Application of fertilizers by (a) broadcasting or by (b) banding in rows.

of nitrogen from a single application of farm manure from large animals is shown in Table 3.1.

Utilizing the data in Table 3.1, if a single application of farm manure delivering 200 lb of N per acre (20 tons of manure per acre) were made, about 100 lb N would be available during the current crop year, with about 16 lb being available in the first following year, 10 lb in the second following year, 4 lb in the third following year, and 2 lb in the fourth and subsequent following years. If the application of 20 tons of manure per acre were made yearly, the residual effects would be additive, as shown in Table 3.2. The benefits of higher or lower applications would be directly proportional to the residual effects of 20 tons per acre. For example after 20 years, from 10

TABLE 3.1
Residual Value of Nitrogen in Solid Farm Manure

Years after Application	Approximate Portion of Nitrogen That Is Available (% of Original Application to Soil)
0*	50
1	8
2	5
3	2
4	1

* Year of application.

TABLE 3.2
Residual Nitrogen after Several Years of Application of Farm Manure at 20 Tons (About 200 lb N) per Acre per Year

Consecutive Years of Application*	Cumulative Nitrogen Available from Previous Applications (lb N/acre)
1	15
2	20
4	26
6	30
8	34
10	38
15	47
20	54

* Years preceding the current crop year.

tons annual manure applications, residual N would be 27 lb per acre; and from 30 tons annual manure applications, residual N would be 81 lb per acre.

Nitrogen-Fixing Legumes

Some crops, such as nitrogen-fixing legumes (for example, soybean, alfalfa, clovers, vetches, and lespedeza), grow in symbiosis with bacteria and obtain their nitrogen from the air and may not require fertilization with nitrogen. Nitrogen-fixing legumes—for example, soybeans and forage legumes—grown as agronomic crops usually are not fertilized with nitrogen, since the symbiotic relationship between the plants and bacteria supply these crops with nitrogen. Garden legumes—peas and beans—usually are fertilized with nitrogen, for often nitrogen fixation by these crops is inadequate to supply them with their nutritional requirements. In general, however, legumes do not respond to nitrogen fertilizer if they are growing symbiotically with the bacteria required for nitrogen fixation. An application of fertilizer nitrogen diminishes the amount of nitrogen that these crops fix symbiotically, roughly offsetting the amount fixed by the amount of nitrogen applied. The bacteria with which the legumes interact to form root nodules, which are the site of nitrogen fixation, must be present in the soil or must be added to the soil for symbiosis and nitrogen fixation to occur. The chapter on green manures presents more information on nitrogen fixation by crops.

Nitrogen Concentrations in Fertilizers

The concentrations of nitrogen in fertilizer range from 1% to 80% of the weight of the material. Naturally occurring organic fertilizers range from 1% to 15% nitrogen. Pure protein is about 16% nitrogen, so materials with concentrations of nitrogen higher than 16% probably are manufactured. A 16% concentration of nitrogen is low for modern-day chemical fertilizers. Organic materials with less than 1% nitrogen generally are too low in nitrogen to be considered fertilizers. These materials are bulky to handle, and they may release nitrogen too slowly to be of value to a crop. In fact, some materials with nitrogen concentrations below 1% will cause depletion (immobilization) of soil-available nitrogen, and additional nitrogen from a fertilizer must be applied to prevent development of deficiencies in crops.

The concentrations of nitrogen in some common organic and chemical fertilizers and in some miscellaneous materials are given in Table 3.3. Although a precise value is given for the concentration and release of nitrogen in the organic materials, the actual composition and release rates of these materials will vary according to the weather and climate, the specific origin of the materials, and how the materials are handled, processed, and stored.

Transformations of Nitrogen in Soil

Plants absorb nitrogen from the soil as nitrate (NO_3^-) or ammonium (NH_4^+) ions. Chemical fertilizers provide these ions by dissolution or by rapid decomposition in the soil:

Dissolution

Ammonium nitrate → Ammonium + Nitrate*

Bacterial decomposition

Urea → Ammonium

**Note:* Ammonium is converted to nitrate in the soil by *nitrification*.

Whereas the decomposition of urea is rapid, the decomposition of organic fertilizers to release nitrogen is relatively slow. The process of release of nitrogen from organic fertilizers to plant-available forms is *nitrogen mineralization* or *ammonification* (Figure 3.6). Rates of mineralization are related to nitrogen concentration in organic fertilizers. Generally, the higher the nitrogen concentration, the faster the rate of mineralization. Practices for uses of organic fertilizers are governed by mineralization of the material. If mineralization is rapid, with 80% or more of the nitrogen being released in one season, nitrogen will be quickly available to plants. If mineralization is slow, 30% or less release of nitrogen per season, plants may be undernourished unless a generous amount of fertilizer is applied.

Rapidly mineralizing organic fertilizers should be handled as if they were chemical fertilizers. Care must be taken to keep quick-release organic fertilizers (blood, seed meals, alfalfa meal) from contact with seeds or young roots. Overapplication of these fertilizers may lead to ammonium toxicity if mineralization proceeds more rapidly than *nitrification* (Figure 3.6).

The residual effects of rapidly mineralizing organic fertilizers, such as dried blood and seed meals, will not be appreciably higher than those of chemical fertilizers and

TABLE 3.3
Concentration of Nitrogen in Common Organic and Chemical Fertilizers and Miscellaneous Materials

Material	Nitrogen (% dry wt)	Release in One Season (% of Total N)*
	Organic Fertilizers	
Feather meal	15	70
Dried blood	12	90
Animal tankage	8	70
Fish scrap	8	70
Soybean meal	7	70
Cottonseed meal	6	70
Linseed meal	6	70
Castor pomace	5	70
Sewage sludge**		
Digested	6	50
Raw	2	10 to 50
Manure, dehydrated		
Large animal	2	30 to 50
Poultry	3	70
Manure, wet		
Large animal	0.5	10 to 50
Poultry	1.5	60
	Chemical Fertilizers	
Anhydrous ammonia	80	100
Urea	45	100
Ammonium nitrate	34	100
Ammonium sulfate	21	100
Diammonium phosphate	20	100
Sodium nitrate	16	100
Calcium nitrate	15	100
Potassium nitrate	13	100
	Miscellaneous Materials	
Compost	1	10 to 50
Spoiled hay		
Grass	1.5	40
Legume	2.5	60
Grass clippings, green	1.5	60
Sawdust, woodchips	0.1	0
Bark	0.3	0
Hides, leather	12	40
Silk, hair, fur	12	30

* Rate of mineralization or availability.
** Not certified organic.

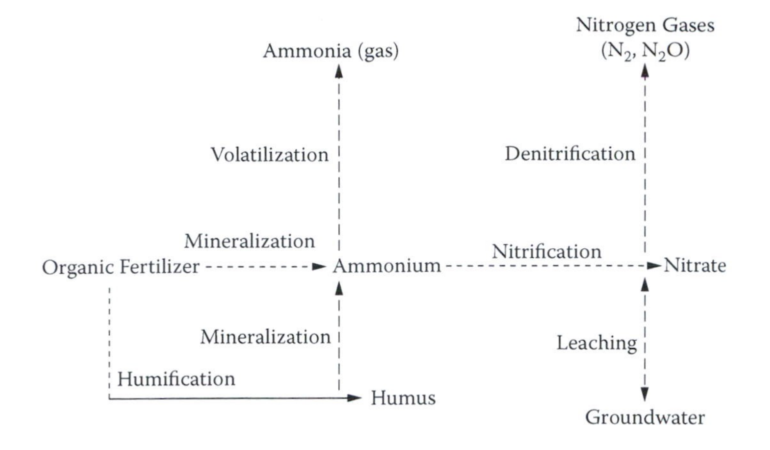

FIGURE 3.6 Transformations of nitrogen in soils (the nitrogen cycle).

will be much lower than those of slowly mineralizing organic fertilizers, such as manures.

Plants may absorb either ammonium or nitrate ions from the soil. However, in soils, nitrate is the prevailing form of mineral nitrogen. Mineralization is a relatively slow process in comparison to nitrification, which proceeds rapidly in well-aerated, warm soils. Consequently, ammonium formed by mineralization is transformed rapidly to nitrate by microbial action. In cold soils, nitrification may proceed slowly so that ammonium may accumulate, particularly if a chemical fertilizer, such as ammonium sulfate or urea, is used. Even under good conditions, dried blood and seed meals may mineralize more rapidly than ammonium is oxidized to nitrate, and ammonium toxicity to seeds or seedlings may result, particularly if fertilization and seeding or planting occur at or nearly at the same time and if care is not taken to keep the seeds or transplants away from the fertilizer. Chicken manures are high in ammonium. The so-called burning of crops by chicken manure is due to poisoning of the plants by ammonium from the manure. It is a wise practice to apply nitrogen-rich organic fertilizers and farm manures about two weeks ahead of seeding or transplanting of crops to avoid ammonium toxicity or other problems that may arise during the rapid decomposition of the materials in the soil.

By a biological process called *humification*, a portion of organic fertilizers or any organic matter incorporated into soil is converted to *humus*. Humus is a relatively stable product of organic matter decomposition in soil. As a rule, organic materials with low nitrogen concentrations will make a larger contribution to humus than organic fertilizers with high nitrogen concentrations. Dried blood, which releases about 80% or more of its nitrogen in a mineral form in one season, will not contribute to increasing the humus in the soil significantly, unless the plant litter that was produced from a crop that grew on the nitrogen from the blood is returned to the soil. However, materials, such as cow manure, that release half or less of their nitrogen in a growing season may contribute substantially to elevating the humus content of soil. Organic matter that does not decompose in one season and is carried over to the next season in a still relatively unstable form is not humus. Also, organic matter incorporated into soil often is decomposed almost completely in two years without a contribution to increasing the humus content of soil. Elevation of humus contents of soil is a long-term process.

Humus although relatively stable is subject to mineralization. About 1% or 2% of the nitrogen in humus in the plow layer (accepted commonly to be 6 inches deep) may be converted to mineral form each year. Available nitrogen from unfertilized soil is directly related to the humus content of soil. However, the rate of mineralization varies greatly with soil conditions—aeration, moisture, temperature—and cannot be predicted easily or with accuracy. Using the mineralization rate of 1% or 2% per year, a fertile mineral soil that has about 4000 lb of nitrogen in the humus in the plow layer may contribute 40 to 80 lb of available nitrogen per acre per year for crop production. A soil with 1000 lb of nitrogen in humus per acre will contribute only 10 to 20 lb of nitrogen or less to a crop. Soil texture and color are fair indicators of organic matter or humus contents in soils. Table 3.4 lists quantities of nitrogen in soils of various colors and textures. The soils represented in this table are types found in considerable areas of the north central states. The amount of nitrogen in soils of various colors and textures will vary with geographic location. In general, nitrogen in soils increases from south to north and from east to west in the United States, because of variations in mineralization due to different temperatures and amounts of water in the soils. Cultivation (growing of crops) leads to depletion of soil organic nitrogen, because of the increased oxidation of humus in the soils with

TABLE 3.4
Quantities of Nitrogen in Soils by Depth, Color, and Texture (lb N/acre)

	Depth (inches)		
Type of Soil	**0 to 6**	**6 to 20**	**20 to 40**
Deep peats	35,000	65,000	98,000
Black clay loams	7200	7500	3200
Brown silt loams	5000	5900	3500
Brown loams	4700	6700	4200
Deep gray silt loams	3600	2200	2300
Brown sandy loams	3100	3900	4200
Yellow gray silt loams	2900	2700	3200
Gray silt loams	2900	3200	3200
Drab silt loams	2800	3200	3400
Yellow fine sandy loams	2200	2610	2700
Yellow silt loams	2000	2000	2400
Light gray silt loams	1900	1900	2100
Sands	1400	2100	3100

Adapted from Soils and Men, *1938 Year of Agriculture*, p. 370, United States Department of Agriculture, Washington, D.C.

aeration imparted by tillage and because of removal of nitrogen by crops. The nitrogen in the top 6 inches of soil should be the portion used to estimate the pool of available nitrogen from the tabulated data (Table 3.4).

Applying the rule that 1 or 2% of the nitrogen in the surface 0-to-6-inch zone would be mineralized gives an estimate of the nitrogen available from each soil. For example, annual releases of nitrogen would be 70 to 140 lb from a black clay and 14 to 28 lb from a sand. The surface zone has the highest concentration of nitrogen, the best aeration, and the highest microbial activity of the three depths. Plants get most of their nutrition from the upper layers of the soil and typically from a zone that is 3 to 10 inches below the surface of the soil. Not much nutrition is acquired from depths below 10 inches. Deep penetration of roots into the soil generally benefits plants in the obtaining of water.

Not all of the nitrogen that is mineralized from organic matter is recovered by plants. Part of it is recycled back to organic matter; some is out of reach of roots, laterally or vertically; and some is lost. About half of the nitrogen released from soil organic matter may be recovered by plants.

Forms of Nitrogen in Soil

Most of the nitrogen in the soil is in organic matter. Humus and other forms of organic matter that are resistant to decomposition are reserves of nitrogen. The amounts in inorganic or mineral forms of nitrogen are small relative to the amounts in organic forms (Table 3.5). Because of *nitrification*, the oxidation of ammonium to nitrate, most of the mineral fraction of nitrogen in soil is nitrate. Nitrate will be in the soil solution (dissolved in the water in the soil), and all nitrate is potentially available to plants. Nitrate is highly mobile in soil and is transported by diffusion in water and by flow of water. Ammonium in soil is usually very low because of the rapid oxidation (nitrification) of ammonium after its release from organic matter. A fraction of the ammonium nitrogen in soil may not be available for absorption by plants because of the retention of ammonium ions by clays. Most terrestrial plants are adapted to soils in which nitrate is the prevailing form of available nitrogen and have low tolerance for ammonium accumulation in soil.

TABLE 3.5
Relative Amounts and Potential Availabilities of Organic and Inorganic Nitrogen Reserves in Soils

Form of Soil Nitrogen	Fraction of Total N in Soil (%)	Potential Availability (%)
Humus-organic matter	98	1 to 2
Nitrate	1.9	100
Ammonium	0.1	95

Immobilization of Nitrogen in Soil

Nitrogen *immobilization* refers to the consumption of soluble or available inorganic nitrogen by soil microorganisms. Immobilized nitrogen is unavailable to plants until the microorganisms die and their organic matter is mineralized. Immobilization occurs if plant residues that are high in carbon and low in nitrogen are incorporated into soil. Layers or mulches of organic matter on the surface of the soil do not cause serious problems from immobilization because contact between the organic matter and the soil is limited. In the soil, carbon in the organic matter is food for certain soil microorganisms, but highly carbonaceous organic matter does not contain sufficient nitrogen to support growth of the microorganisms. Consequently, the microorganisms use the available mineral nitrogen in the soil for their growth. Since microorganisms proliferate throughout the soil much better than plant roots and are more intimately in contact with soil nitrogen than roots, microorganisms have a better capacity to obtain nitrogen than plants.

Immobilization of other nutrients is possible, particularly phosphorus and sulfur, but immobilization is most apparent with nitrogen. Organic materials that can cause immobilization are paper, sawdust, wood chips, bark chips, straw, tree leaves and needles, coarse dead garden residues, mature corn stalks, and other plant residues that are low in nitrogen. Generally, residues with less than 1% nitrogen by weight have the potential to cause immobilization in soil. Many growers will not incorporate crop residues in which the amount of carbon exceeds nitrogen by a ratio of 35 to 1 (C:N ratio = 35:1) but will leave these materials on the surface of the ground or remove them from the land. Residues that have carbon to nitrogen rations far greater than 35:1, for example, 100:1, should be composted before they are worked into soils. If these residues should be worked into soils, to avoid nitrogen-deficient crops, additional nitrogen must be supplied as fertilizer to compensate for the nitrogen that will be immobilized. About 1 lb of nitrogen should be added for every 100 lb of residues that are incorporated with carbon to nitrogen ratios greater than 35:1. Carbon to nitrogen ratios of some common organic materials are given in Table 3.6.

TABLE 3.6
Carbon to Nitrogen Ratios of Organic Materials

Organic Matter	Nitrogen (% dry wt)	Carbon:Nitrogen
Paper	0.01	5000
Sawdust	0.1	500
Wood chips	0.1	500
Bark chips	0.2	250
Straw	0.3	180
Peat moss	0.5	100
Cow manure	2.0	25
Grass hay	1.5	33
Alfalfa hay	2.5	20
Seed meals	6.0	8

If sawdust or wood chips are worked into soil, nitrogen immobilization will occur for up to 3 or 4 years after incorporation. Sawdust will give more serious immobilization than wood chips, because sawdust is more finely divided. Paper will cause intense immobilization, but its effect will be of shorter duration than that of sawdust since the paper will decompose more rapidly than sawdust. The immobilization effects of straw and other coarse crop residues will last for about 1 year. Some materials, even though they have wide ratios of carbon to nitrogen, will not cause immobilization. These materials are those that are stable and resistant to decay. Plastic will not cause immobilization because of its resistance to microbial attack. Peat moss will not stimulate immobilization, for peat moss is a stable form of organic matter, being mostly structural materials (lignin) remaining after the readily decomposable materials have removed by microbial action. Peat moss can be added to potting media and garden soils without dangers of nitrogen immobilization.

Organic materials that have carbon to nitrogen ratios slightly greater that 35:1 initially will stimulate immobilization, but as the ratios narrow by rotting of the organic materials these materials will release nitrogen by mineralization. However, since the period of immobilization may occur at the time when young crops need nutrition of nitrogen, growers should make an application of nitrogen to compensate for immobilization. If the application of organic matter can be made 6 to 8 weeks in advance of planting of the crop and if weather permits rotting of the organic matter in the soil, a grower might be able to omit or use only light applications of nitrogen with materials with carbon to nitrogen ratios of less than 100.

Organic materials with C:N of 20:1 to 35:1 can be added without supplemental nitrogen fertilizer and with only a 1- or 2-week wait between application of materials and planting. This 1- or 2-week wait is desirable as it permits for some stabilization of organic matter in the soil before plants and allows problems of rotting diseases and toxicities of ammonia or other organic substances to dissipate before planting of crops. Organic materials with very narrow ratios, such as those less than 10:1 to 15:1, are fertilizer-grade materials and should be treated as such and not as materials to increase the organic matter content of soils. A wait of 1 or 2 weeks between application of organic fertilizers and planting is recommended with organic materials with narrow C:N ratios. With these materials, rapid mineralization soon after their application may elevate ammonium levels in soil and injure seeds and seedlings. After 2 weeks, most of the ammonium will be oxidized to nitrate. Also, as with manures and seed meals, rot diseases of plants will be stimulated by additions of organic matter. The organisms that are rotting the manures or seed meals will rot seeds and plant roots as well as the organic fertilizers. A 1- or 2-week delay between fertilizing and planting allows for the proliferated population of microorganisms to die to noninjurious levels.

PHOSPHORUS

Functions

The discovery of the essentiality of phosphorus is attributed often to the German physical chemist Justus von Liebig for his work around 1840 in field experiments

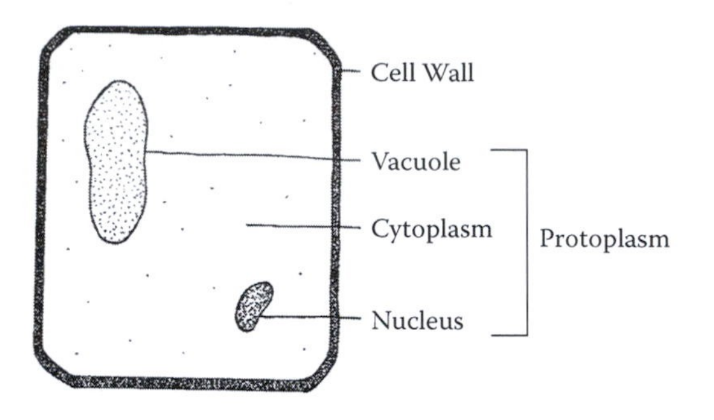

FIGURE 3.7 Diagram of a plant cell, illustrating the nucleus in which phosphorus is concentrated.

with phosphorus-containing materials (bones and bones treated with sulfuric acid to produce superphosphate) and with his development of chemical fertilizers (known then as *patent manures*). Definitive proof of the essentiality of phosphorus came from experiments by other German researchers (Wilhelm Knop, Julius von Sachs, W.F.K.A. Salm-Horstmar), who around 1860 conducted research with solution culture (hydroponics) of plants. With solution culture, these researchers were able to add or withhold elements from the medium to demonstrate or confirm the essentiality of all of the macronutrients and iron. The nutrient solutions at that time were not sufficiently pure to demonstrate the essentiality of micronutrients other than iron.

Phosphorus is concentrated in the nucleus of cells (Figure 3.7). In the nucleus, phosphorus is largely in the genetic material (DNA). Outside the nucleus, the ribonucleic acid (RNA) is a major phosphorus-containing constituent, with roles in protein synthesis. Energy transformations in cells require phosphorus. Phosphorus is a component of adenosine triphosphate (ATP), which is a product of energy transformations or transfers during photosynthesis and respiration. In addition, phosphorus is a component of lipids, carbohydrates, proteins, enzymes, coenzymes, and other plant metabolites. In plant bodies, phosphorus is concentrated in the growing tips of shoots and roots where cell division occurs (Figure 3.8). Phosphorus may be 100 times more concentrated in dividing cells than in surrounding cells.

Effects of Phosphorus on Plant Growth and Quality

Phosphorus fertilization of plants tends to balance some of the effects that may occur with nitrogen fertilization. Phosphorus fertilization increases the strength of stems by increasing the thickening of cell walls. The strengthened cell walls increase the resistance of plants to diseases. The rate of maturation of crops is accelerated by phosphorus fertilization, thereby somewhat, but not fully, offsetting the delay in maturity that may occur with nitrogen fertilization. Lateral root development (branching) is stimulated by phosphorus. Roots proliferate in phosphorus-rich zones of soil due to the high availability of phosphorus and its stimulation of cell division. Lengths of primary roots may not be changed by phosphorus fertilization as the effects are principally on the development of branch roots.

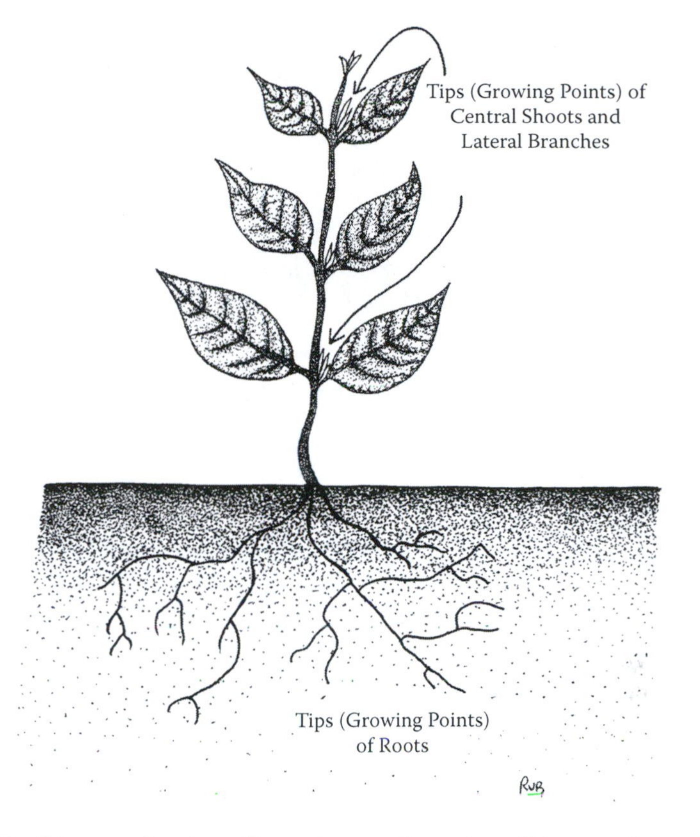

FIGURE 3.8 Diagram of a plant, illustrating growing points of shoots and roots in which phosphorus is concentrated.

Phosphorus is essential for flowering of plants. If phosphorus is deficient, flower size and abundance will be suppressed. Supplying of phosphorus to a deficient plant will increase flowering, but once sufficiency of phosphorus has been achieved, further increases in phosphorus fertilization will not increase flowering. If flowering has been restricted by an overabundance of nitrogen supply, phosphorus fertilization generally will not reverse this effect. Time is needed for plants to enter the reproductive stage and to begin flowering.

Symptoms of Phosphorus Deficiency in Plants

Phosphorus-deficient plants grow slowly. In early stages of deficiency, they are dark green, often gray-green or blue-green. These dark-green or off-green colors are the result of stunted growth in which the green pigment is concentrated in leaves that remain small. Diagnosis of phosphorus deficiency by recognizing stunting is difficult, unless a well-nourished plant is available as a reference, because except for size, the stunted, deficient plant often appears normal. A more striking characteristic occurring with the advancement of deficiency is a purpling of the leaves. The older leaves, even those of young seedlings, appear purple, particularly on the leaf undersides. As deficiency becomes more severe, the lower leaves become yellow, become brown, and then drop off. Not all plants develop these symptoms; for example, with cucumber, plants are stunted, but the lower leaves do not develop the off-green or red colors. Instead, the tissues die at the base of the leaves near the petioles or along the margins of the leaves. These tissues dry and crumble away as the deficiency progresses.

Phosphorus is a mobile element in plants. If the supply of phosphorus in soil is exhausted, plants will transport phosphorus from the lower leaves to the growing regions or to the flowers, fruits, and seeds. Thus, the symptoms are most prevalent on the lower, old leaves of the plants.

Cold weather increases the likelihood of phosphorus deficiency. In cold soils, phosphorus is sparingly soluble, and in addition, its absorption by plants is slowed. Young seedlings started in flats indoors or outdoors often become chilled. These seedlings appear stunted and purple. Warming the flats, applying phosphorus fertilizers, or taking both actions usually overcomes the deficiency. Plants that are transplanted outdoors often show symptoms of phosphorus deficiency in cold weather. Early plantings of crops in the field or garden should be well-fertilized with phosphorus. In the early spring, soils are cold, plant roots are sparse, and phosphorus is slowly mobile in the soil. These phenomena restrict the availability of phosphorus to plants. An abundant supply of phosphorus is necessary to ensure that the young crop gets ample nutrition. Hence, it is important that all of the phosphorus fertilization of crops occurs at planting. This practice ensures that an abundant, concentrated supply of phosphorus is available to the young plants with sparse roots in cold soils. Also, phosphorus is fixed readily in soils so that it becomes more unavailable to plants with time in the soil. If phosphorus is placed far from roots, it may be fixed before the roots grow into the zone of fertilization. Contrary to some beliefs, the longer phosphorus fertilizers are in the soil, the less available the phosphorus becomes to plants.

Phosphorus deficiency in a plant is difficult to rectify. It is difficult to get phosphorus into a phosphorus-deficient plant. Phosphorus mobility in soils is restricted by the chemistry of phosphorus in soil. Phosphorus may move only a few centimeters from its point of application. Surface applications of phosphorus may not reach the roots of plants already growing in the soil. However, if phosphorus is readily available early in the growing season, a plant will absorb phosphorus in excess of its current needs. Later in plant development, this abundance of phosphorus can be mobilized to regions of the plant with demands for it.

Amounts of Phosphorus Required by Crops

Crops differ in the amounts of phosphorus that they remove from the soil. As with nitrogen, amounts of removal are related to the productivity of crops. High-yielding, rapid-growing, densely planted crops remove more phosphorus than low-yielding, slow-growing, sparsely planted crops. Phosphorus removal is about one-fourth that of nitrogen. Phosphorus concentrations in leaves of field-grown crops with adequate nutrition range from 0.2% to 0.5% of leaf dry weight. Concentrations below 0.15% usually indicate phosphorus-deficient conditions. Concentrations above 0.5% occur in plants grown in hydroponics, in well-fertilized soils, or in organic media. Some approximations of phosphorus consumption by plants according to yield potential or plant growth follow.

High Demand for Phosphorus

More than 30 lb actual phosphorus removal per acre (0.75 lb P per 1000 sq ft). High-yielding, densely planted, rapidly growing crops:

Corn
Celery
Potatoes
Tomatoes
Cucumbers
Forages
Cotton
Cereal grains (high yielding)

Moderate Demand for Phosphorus

Between 15 and 30 lb actual phosphorus removal per acre (0.38 to 0.75 lb P per 1000 sq ft). Moderate-yielding crops:

Most vegetable crops
Soybeans
Grain sorghum
Cereal grains (low or moderate yields)

Low Demand for Phosphorus

Less than 15 lb actual phosphorus per acre (0.38 lb P per 1000 sq ft). Low-yielding, slow-growing, sparsely planted crops:

Peppers
Orchards
Beans

Fertilization with phosphorus moderately in excess of the needs of a crop generally does little harm to a crop. Luxury consumption, known as the accumulation far beyond the needs of a crop without any effect on growth, of phosphorus rarely occurs. However, in some crops, excessive phosphorus in soils has been associated with occurrence of deficiencies, such as zinc or iron. It is suggested that high levels of phosphorus in soils may hold zinc or iron in sparingly soluble complexes and restrict the availability of these nutrients to crops. High accumulations of phosphorus in plants have been associated with reduced metabolic availability of zinc or iron and with the appearance of deficiency symptoms for zinc or iron.

Due to the low amount of leaching of phosphorus in soils, phosphorus can accumulate with repeated applications of fertilizers to land. If soil tests reveal *high* concentrations of available phosphorus, growers should consider not adding any additional phosphorus fertilizers for that cropping season at least. In sandy soils, applications of phosphorus beyond agronomic rates, especially with farm manures, have been associated with leaching of phosphorus into the substrata of soils or into groundwater. In fine-textured soils (clayey or silty soils), leaching may be nil, but losses may occur by surface runoff and erosion. Erosion normally carries away the fine particles of soils more than the coarse particles, thereby leading to the potential for substantial transport of phosphorus from the land. This loss of fine particles not only depletes land of phosphorus, but the particles may enter bodies of water and enrich the sediment with phosphorus, thereby promoting the growth of plants and algae in the water.

Phosphorus Concentrations in Fertilizers

The concentrations of phosphorus in fertilizers range from about 1 to over 50% expressed as oxide (P_2O_5). Materials that contain less than 1% P_2O_5 may be too low in phosphorus to be of value as a fertilizer for a single application. The phosphorus concentrations of some common organic and chemical fertilizers, expressed as P_2O_5 and as actual P, are given in Table 3.7.

Rock phosphate is an ore of marine biological origin. The United States has large deposits of rock phosphate, making the United States one of the foremost producers

TABLE 3.7
Phosphorus Concentrations in Common Fertilizers

Fertilizer	Phosphorus Concentration (%)		Relative Availability*
	P_2O_5	P**	
	Organic		
Rock phosphate	30	13	Very low
Colloidal rock phosphate	20	9	Very low
Basic slag	10	4	Moderate
Bonemeal	23	10	Moderate
Composts	1	0.4	Moderate
Manures, dehydrated	1	0.4	Moderate
	Chemical		
Ordinary superphosphate	20	9	High
Triple superphosphate	45	19	High
Diammonium phosphate	52	22	Water-soluble

* Relative release of phosphorus from fertilizer by dissolution or by mineralization.
** Actual P; 0.44 × P_2O_5.

of phosphorus fertilizers in the world. China, west Africa, and other world areas have large deposits of rock phosphate and are current or potential major producers of phosphate fertilizers. Rock phosphate has been mined in the United States since about 1867, starting with deposits in South Carolina. Currently, phosphate for application to land is largely from deposits in Florida, accounting for about 87% of phosphate mining in the country. North Carolina, Idaho, and Utah also have active mines.

Rock phosphate as such has very low solubility in water. It is chemically similar to the enamel of teeth. Although rock phosphate is concentrated, around 30% total P_2O_5, the available P_2O_5 is only about 3% or less of the mass. Rock phosphate is the source from which most chemical phosphorus-containing fertilizers are made by treatment of the ore with acids to create superphosphate.

Colloidal rock phosphate is a lower grade of fertilizer than ordinary natural rock phosphate. Some of the ordinary rock phosphate mines have been depleted or have become uneconomic to operate so that colloidal rock phosphate from Florida is the dominate source of rock phosphate in the marketplace.

After rock phosphate arrives at a fertilizer plant, it is processed in water to remove sand and clay. Sand is washed away by a spiral of water, and a slurry of rock phosphate and clay remains. Clay is removed by allowing it to rise in settling tanks where it is skimmed from the surface of the water, or the slurry also may be allowed to settle in lagoons. Fine particles of rock phosphate adhere to the clay. This mixture is called *colloidal rock phosphate*. The term *colloidal* refers to the clays in the product and does not imply that the material is superior to ordinary natural rock phosphate because of a fine particle size. In fact, the clay dilutes the concentration of phosphorus and does not improve its availability. Colloidal rock phosphate has about 20% total P_2O_5 compared to about 30% total P_2O_5 in ordinary rock phosphate. Colloidal rock phosphate is sometimes referred to as *soft rock phosphate*. This source is not any softer than ordinary rock phosphate as both have the same chemical composition in the basic phosphate-carrying compounds.

Considerable attention must be given to use of ordinary or colloidal rock phosphate to ensure its release of phosphorus. The following steps must be followed in the use of rock phosphate; otherwise, applications of rock phosphate to land will have little or no agricultural benefit.

1. *The rock must be finely ground and mixed thoroughly with the soil.* Silt-sized particles of rock are necessary to give close contact between soil and the rock and to increase the weathering of rock in the soil. If the particles are sand sized, little phosphorus will dissolve from the fertilizer, and plant nutrition will be poor.
2. *The soil should be acid.* A pH 5.5 is recommended. Acid helps to dissolve the rock, thereby making its phosphorus more available to plants. The acid in soils converts rock into a form similar to superphosphate. In more strongly acid soils, however, iron and aluminum may be so high in solution that they precipitate phosphates from the soil solution, rendering the phosphorus unavailable to plants before it can be transported into a zone in which the plant roots are growing. And, plants roots that grow into the zone of fixed phosphorus will not be able to absorb the fixed phosphorus.

3. *Organic matter should be added with the rock phosphate.* Any kind of organic matter of plant origin seems to work to increase the solubility of rock phosphate. Decaying organic matter produces acids that help to dissolve rock phosphate. Organic matter forms chelates (forms organic complexes) with iron and aluminum so that iron and aluminum do not react with phosphates to form sparingly soluble precipitates and thus helps to keep the phosphorus in solution. Organic matter should be added at about 1 lb per square foot of land. Rock phosphate can be mixed with farm manures—about 50 lb of rock phosphate per ton of manure. The rock fortifies the manure with phosphorus, and the acids in the manure facilitate the dissolution of the rock. Adding the rock phosphate and manure to land together helps to keep the phosphorus available to plants.
4. *Rock phosphate should be applied in amounts that are two, four, or even up to 10 times the recommendations for phosphorus application indicated by soil tests for chemical fertilizers to allow for the low availability of phosphorus in rock phosphate.* The availability of phosphorus from rock phosphate is so low that higher than recommended applications are needed to ensure adequate plant nutrition. For example, although the total P_2O_5 in rock phosphate might be 30% by weight, only about 3% of the mass is available P_2O_5. For colloidal rock phosphate, the available P_2O_5 is about 2%. Rock phosphate may be a relatively inexpensive source of phosphorus, but the requirement that it be applied in large amounts may eliminate the benefits of the cheap price.

Basic slag is a by-product of cast iron and steel production from phosphorus-containing iron ores. The procedure for making steel and cast iron from these ores was developed in England in about 1877. Phosphorus is removed from molten iron ore during the manufacture of cast iron and steel, because presence of phosphorus weakens these products. To remove the phosphorus, limestone and iron ore are melted in an open heater. Limestone reacts with phosphorus in the ore, and this product being lighter than iron floats to the top of the molten mass where it can be poured off as slag. The cooled slag is ground to a fine powder to produce the fertilizer. Basic slag is an excellent fertilizer of almost equal value to ordinary superphosphate. Basic slag is a scarce product, usually available only in the vicinity of steel mills that use an open heater system. Many steel mills are stockpiling the slag on their properties and are not offering the material for sale.

Basic slag is an alkaline material, having a liming equivalence of about 70% of that of agricultural limestone. Basic slag can be applied by mixing with the soil or by banded applications along the rows of crops at planting. Organic matter increases the availability of phosphorus from slag. Soil pH should be about 6.0.

Bonemeal is almost as effective as ordinary superphosphate as a source of phosphorus. The value of bones as fertilizer has been recognized for almost as long as people have been farming. The agricultural benefits of bones were first attributed to the fats and gelatin in the bones. Phosphorus was discovered as an element in about 1669 in Germany, and over the next 100 years, phosphorus was shown to be a constituent of bones. It was recognized then that the value of bones in agriculture was in their phosphorus content. Bones as a fertilizer became so popular in England that the home supply was soon exhausted, and bones were imported from any available sources.

Bonemeal is produced by boiling and steaming raw bones to remove fats and then grinding the bones. Most of the bones are from abattoirs. Bones are ground finely to make the fertilizer. Steamed bonemeal is almost free of nitrogen. Raw bonemeal has about 4% nitrogen, but should be avoided for sanitary reasons. Most of the marketed bonemeal is steamed. Bonemeal can be applied to soil by mixing or banding. Organic matter also helps improve the availability of its phosphorus. Soil pH should be about 6.0 to limit phosphorus fixation.

Plant-derived materials, such as *composts* and *manures*, are too low in phosphorus to be used alone as phosphorus fertilizers on poor soils. After phosphorus levels in soil have been built up by fertilization, plant materials can be used to maintain phosphorus fertility in soil. Composts and manures often are enriched in phosphorus by mixing them with rock phosphate or bonemeal incorporated at about 50 lb of fertilizer per ton of compost or manure. Long-term applications of farm manures at rates above the agronomic needs to meet the phosphorus requirements of crops can lead to phosphorus accumulation in soils. On sandy soils, phosphorus might leach into the soil profile and into ground water under circumstances that lead to high levels of phosphorus accumulation. The organic phosphorus in manures and composts may be more mobile in soil than phosphorus from mineral sources.

Superphosphates are manufactured by treating rock phosphate with acids. Acid treatment produces compounds from which phosphorus is more available than those in rock phosphate but not necessarily more concentrated. Liebig experimented with treatment of bones with sulfuric acid around 1840. In 1842, John B. Lawes and Joseph H. Gilbert of the Rothamsted Experiment Station in England secured a patent for the process. Subsequently, treatment of rock phosphate with sulfuric acid led to the development of the superphosphate industry. The same actions imparted by sulfuric acid occur naturally in acid soils and in composts or manures, but the process is much slower and much less complete than that which occurs in the manufacture of superphosphate. Because rock phosphate is treated chemically with sulfuric acid, superphosphate is not considered to be an organic fertilizer.

Ordinary superphosphate is produced by treatment of rock phosphate with sulfuric acid. Triple superphosphate, sometimes called *concentrated superphosphate*, is manufactured by treatment of rock phosphate with phosphoric acid, which is manufactured also from rock phosphate. Among commercial fertilizers, triple superphosphate is the dominant material because of its high concentration of phosphorus and relatively lower shipping costs per unit of phosphorus.

Ammoniated phosphates are made by reacting superphosphates with ammonium. *Diammonium phosphate* is readily soluble in water. It is a constituent of many concentrated water-soluble fertilizers used in greenhouse crop production and in fertilizer solutions. Common fertilizers for houseplants contain diammonium phosphate.

Recovery of Phosphorus from Fertilizers

Less than 20% of the available phosphorus from fertilizers is recovered by crops in the first year following the application. An even smaller portion is recovered by the

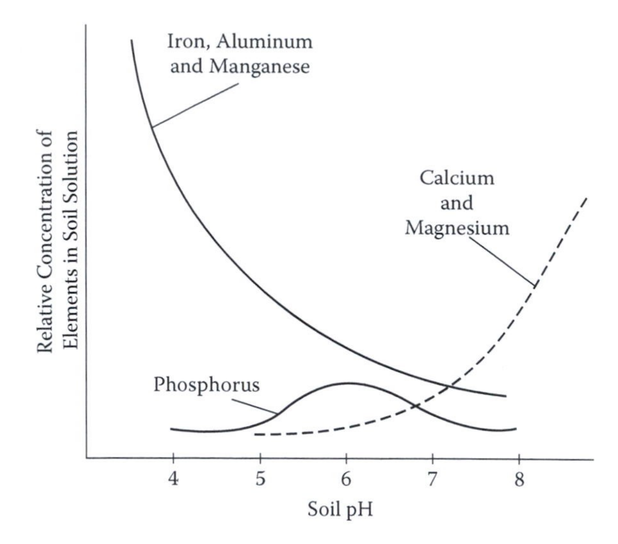

FIGURE 3.9 Potential changes in concentrations of phosphorus, iron and aluminum, and calcium and magnesium in soil solutions, as a function of soil acidity (pH).

succeeding crops, perhaps only 4% in the second year and less than 1% afterward, assuming that 20% is recovered in the first year. Recovery in the first year may be as low as 3% in some soils. Phosphorus is *fixed* in mineral soils by inorganic aluminum, iron, manganese, and calcium or magnesium, which react with phosphate to form sparingly soluble compounds. Fixation occurs in a matter of a few days. The greatest availability of phosphorus in soils occurs at about pH 6.5, or in a range from 6 to 7, at which the capacity of the soil to fix phosphorus is lowest. At this pH, soluble aluminum, iron, manganese, and calcium are in relatively low concentrations, and the fixation of phosphorus by oxides is reduced greatly.

In acid mineral soils, below about pH 5.5, soluble aluminum, iron, and manganese ions increase in the soil solution (Figure 3.9). These ions react with soluble phosphates to precipitate them immediately, eventually rendering the phosphorus unavailable to plants. Also in acid soils, phosphorus is fixed by oxides of aluminum, iron, and manganese. These oxides coat clays and other particles (Figure 3.10). Total phosphorus fixation by these coatings may exceed that which occurs by soluble aluminum, iron, and manganese. Fixation by oxides may occur in soils of neutral pH.

Above pH 6.5 to 7, fixation by calcium increases sharply. In alkaline soils, calcium carbonate (lime) is the prevailing cause of the basicity. In these soils, phosphates and calcium react to form sparingly soluble calcium phosphates. These phosphates generally are soluble enough to meet the phosphorus requirements of plants, whereas those of iron and aluminum are not.

Newly precipitated phosphates of aluminum, iron, manganese, and calcium are finely divided. New precipitates of phosphates on oxides reside on the surface of the particles. In each case, the surface areas of the precipitated phosphates are high so that the availability of the precipitated phosphates might be sufficient to support crop growth. With time, however, changes occur in the precipitates so that the phosphorus

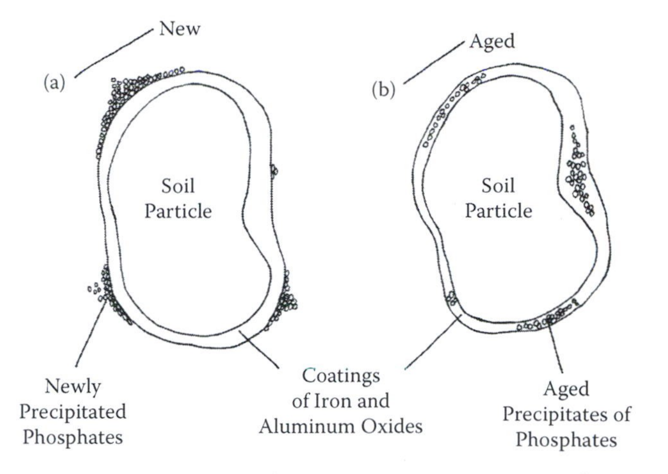

FIGURE 3.10 Representation of (a) new precipitates of phosphates on surfaces of soil particles and of (b) aged precipitates that have hardened or migrated into the surface coatings of the particles.

becomes less available than it was in the original precipitates. Precipitated compounds form larger crystals, thereby reducing their surface area and the availability of the precipitated phosphorus (Figure 3.10). Phosphates that were precipitated on the surface of oxides migrate more deeply into the oxides, leaving less of the phosphorus near the surface of the particles and reducing its availability. The longer that phosphorus is in the soil, the less available it becomes because of its migration into the oxides and formation of large crystals. Contrary to some belief, no advantage is gained by applying phosphorus well in advance of planting of crops, since phosphorus becomes less available the longer it is in the soil. All of the phosphorus fertilization of crops should occur at planting of the crops.

Increasing the Availability of Soil and Fertilizer Phosphorus

The capacity of acid soils to fix phosphorus can be reduced by liming of acid soils. Liming of a soil at pH 4 to pH 6.5 may reduce phosphorus fixation by half. This effect may be expressed in slower fixation of phosphorus from fertilizers and in a reduction in the total capacity of the soil to fix phosphorus. The increased availability of phosphorus in limed soils is due in large part to the reduction in soluble aluminum and iron in solution (Figures 3.9 and 3.11). Limed soils still have enormous capacities to fix phosphorus, and additional practices may need to be taken to improve the availability of phosphorus in mineral soils.

Addition of organic matter increases the availability of phosphorus. The products of decaying organic matter form complexes, called *chelates*, with ions of aluminum, iron, and manganese to bind or sequester these ions so that their reactivity with phosphates is lessened (Figure 3.12). Consequently, more phosphorus remains in solution in the presence of organic matter than in its absence. Organic acids that are produced during the decay of organic matter also facilitate the dissolution of

Liming of Soils

Soluble Iron and Aluminum in Soil [Al^{+++} and Fe^{+++}]	+	OH^-	-------→	Precipitated Iron and Aluminum [$Fe(OH)_3$ and $Al(OH)_3$]

FIGURE 3.11 Removal of iron and aluminum by precipitation from the soil solution by liming of the soil.

difficultly soluble phosphates, essentially converting the difficultly soluble material into superphosphate. This action in combination with the binding of aluminum and iron increases the availability of soil-borne phosphorus and slowly sparingly materials such as rock.

Application of organic matter in combination with liming may further increase the availability of phosphorus from fertilizers above the application of either material alone. Liming decreases the soluble aluminum, iron, and manganese, and organic matter complexes some of the ions that enter solution so that the fixation capacity of the soil is reduced markedly.

Phosphorus from organic matter behaves in soil much as that from inorganic matter so that organic fertilizers, such as manures and composts, are more beneficial in limed soils than in strongly acid soils.

Rock phosphate, although requiring the acids of soils to dissolve, does not work well in strongly acid soils. In soils below pH 5, the concentrations of aluminum and iron are so high that any phosphorus that is dissolved from the rock is precipitated before it can move far enough in the soil for plant roots to obtain it.

The phosphorus from a granule of chemical fertilizers moves only a short distance in soil, less than a centimeter from the granule, before being precipitated. To prevent rapid reaction of phosphorus with the soil, chemical fertilizers—and even bonemeal or basic slag—are placed in bands localized near the roots of plants (Figure 3.13).

(a) Acid Soils				
Aluminum and iron ions (Al^{3+} & Fe^{3+})	+	Phosphate ions ($H_2PO_4^-$ and others)	→	Sparingly soluble AlFePhosphates (unavailable P)
(b) Acid Soils Amended with Organic Matter				
Aluminum and iron ions (Al^{3+} & Fe^{3+})	+	Organic matter	→	Chelated Al^{3+} & Fe^{3+})
Chelated Al^{3+} & Fe^{3+})	+	Phosphate ions ($H_2PO_4^-$ and others)	→	Limited reactivity of Al^{3+} & Fe^{3+}) with phosphates

FIGURE 3.12 Reactions of iron and aluminum in acid soils (a) without organic matter added and (b) with organic matter added.

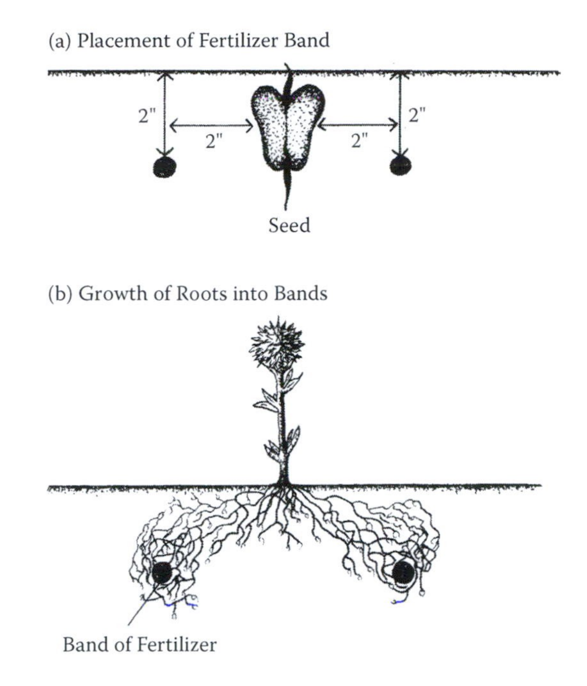

FIGURE 3.13 Diagrams of placement of fertilizers (a) in bands 2 inches to the side and 2 inches below seeds (or transplanted seedlings) and (B) proliferation of roots near the bands of fertilizer.

Typically, these bands are placed about 2 inches to the side and 2 inches below the row of crops or seeds at planting. Modern chemical fertilizers are pelleted or granulated so that they flow better and do not cake. Furthermore, the pellets or granules reduce the contact of phosphorus with soil so that fixation is lessened. These fertilizers need not be applied in bands. Rock phosphate should be ground finely and should not be placed in bands. Rock phosphate is so sparingly soluble that it should be mixed well in the soil to increase its reactivity through contact with the acids of the soil.

Fate of Fixed Phosphorus

Phosphorus that is fixed usually is not lost from the soil. It does not leach into ground water but remains in place in the soil except in sands or in soils that are very low in aluminum and iron. Unless erosion takes the topsoil, heavily fertilized soils will retain their phosphorus. An acre of naturally fertile soil may have 1600 lb phosphorus in the top 6 inches. After years of fertilization phosphorus may build up in excess of 4000 lb per acre. This phosphorus is slowly available. If the reserve from past additions is large enough, further fertilization may be unneeded, or only small maintenance applications of phosphorus will be needed. Soil tests give a good assessment of the fertility of a soil with respect to phosphorus and provide a good basis for recommendations for applications of phosphorus fertilizers.

TABLE 3.8
Phosphorus Contents in the Top 6 Inches of an Acre of Soil

Great Soil Group	Soil Order	Avg lb P per Acre	Location Sampled
Podzols	Spodosol	620	MA, ME, MI, MN, PA
Gray-brown podzols	Alfisol	1200	VA, DC, IN, OH, KY, MD
Red and yellow	Ultisol	800	MS, AL, NC, GA, FL, TX, SC
Prairie	Mollisol	1140	NE, IA, KA, MO
Chernozem northern	Mollisol	1320	SD, NE, WE
Chernozem southern	Mollisol	700	TX
Dark brown and brown	Mollisol	1320	ND, NE, SD, WY
Desert	Aridisol	1740	ID, NE, UT, OR

Adapted from Soils and Men, *1938 Yearbook of Agriculture*, p. 381, United States Department of Agriculture, Washington, DC.

Some soils are naturally more fertile in phosphorus than others, but the range of phosphorus concentrations in unfertilized soils is not large (Table 3.8). Only a small amount of the phosphorus in the plow zone (top 6 inches) of a soil is available at any one time, normally less than 1 lb per acre. In the plow zone, one-third to one-half of the phosphorus is held in the organic matter. In the subsoil, below the plow zone, less than one-fifth of the phosphorus is in organic matter.

POTASSIUM

Functions

The discovery of the essentiality of potassium is credited to researchers (Sachs, Knop) who grew plants in solution culture in the 1860s. However, almost 100 years passed before the functions of potassium in plants were defined. Although potassium is accumulated in plants in amounts that are usually second only to nitrogen among the soil-derived nutrients, its roles in plant metabolism were difficult to identify. Potassium is held by ionic bonds to organic compounds, but unlike nitrogen and phosphorus, it is not a covalent constituent of any organic compound in plants, meaning essentially that potassium is not a component of any plant metabolite. Because of the difficulty of identifying a direct role of potassium in plant metabolism, its essentiality has been associated with many processes. The requirement for potassium has been associated with syntheses of chlorophyll, complex carbohydrates, and proteins. Potassium is now known to be essential for protein synthesis, although it is not a constituent of proteins. The effects on syntheses of chlorophyll and carbohydrates apparently are indirect and are associated with the roles of potassium in protein synthesis. Potassium is needed for the activation of several enzymes in plants. Enzymes are organic catalysts in plant metabolism, and potassium is needed for the maintenance of these catalysts in the proper structural configuration for their activity in plant metabolism (Figure 3.14).

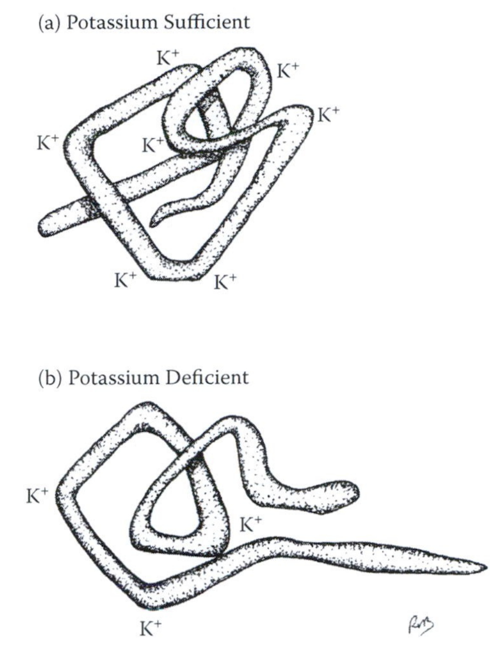

FIGURE 3.14 Concepts of (a) enzyme (protein) surrounded by potassium ions (K^+) and maintained in an active, folded configuration and (b) enzyme deficient in potassium and lacking an active configuration.

Potassium is needed for plant growth by cell enlargement (Figure 3.15). Potassium accumulation in the cell sap promotes water movement into cells. The water in the cells causes forces called *turgor pressure* that keep the cells turgid. The turgor pressure of the water on cell walls enables stretching of the cells and growth by enlargement. Potassium has roles in governing gas exchange by plants. Gases move in and out of leaves through openings called *stomates* in the leaf epidermis. Through imparting changes in turgor pressure in *guard cells* surrounding the stomates, potassium controls the size of the openings and governs rates of diffusion of oxygen, carbon dioxide, and water vapor into and from leaves.

Effects of Potassium on Plant Growth and Quality

Potassium is required for development of fruits and seeds (Figure 3.16). Yields of fruits and seeds will be limited by potassium deficiency, due to smaller sizes of fruits and seeds, poor filling of seeds, and poor quality of produce (culls). Ripening may occur prematurely. Fruits of potassium-deficient plants may be misshapen. Seeds fail to develop fully during potassium deficiency. Grains of potassium-deficient wheat and other cereal grains will be chaffy or unfilled, whereas those of potassium-sufficient plants will be plump with high test weights and with high carbohydrate contents. Ears of corn may be only partly filled, with kernels forming near the base of the ear but not near the tips. These poorly formed ears of corn are called nubbins

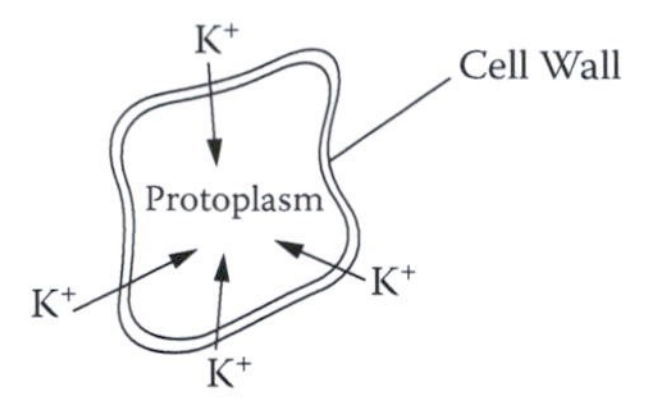

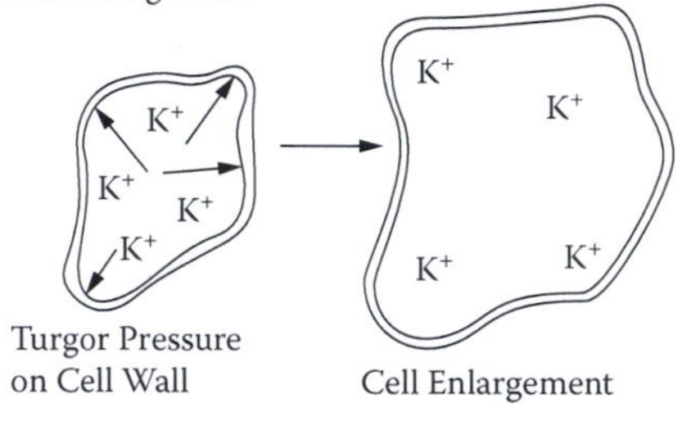

FIGURE 3.15 The role of potassium in plant growth by cell enlargement resulting from potassium ions and water entering into cells to produce turgor pressure as a force for cell expansion.

or owl heads. Adequate potassium is required for all root and tuber crops. In deficient crops, root and tuber development are poor. These organs may be spindly or not fleshy, have tendencies to rot, and appear dark colored. The sugar contents of fruits are increased by potassium fertilization. The sugar contents of sugar beets and sugarcane increase with potassium fertilization. The requirements of potassium for proper development of fruits, tubers, and roots led early investigators to suggest that potassium had roles in carbohydrate metabolism in plants.

Stems of potassium-deficient plants are weak or brittle. Lodging (irreversible falling over) of plants is intensive in potassium-deficient corn and small grains. Stems of potassium-deficient plants are weak because cell walls do not thicken by lignification, and the fibers and supporting elements of stems lack strength. The pith and cortex of stems tend to break up during potassium deficiency, further weakening the support of the plant. Prop roots of corn weakened by potassium deficiency lend little support to the stalks. Weakened, potassium-deficient plants are sensitive to infestation by diseases. Fruits of potassium-deficient plants may not store well. Nitrogen fertilization of potassium-deficient crops increases incidences of diseases, particularly rots.

Roots of potassium-sufficient plants permeate into a larger soil volume than those of deficient plants. Plants well supplied with potassium, therefore, are more resistant to drought than deficient plants.

Symptoms of Potassium Deficiency in Plants

Potassium is a mobile element in plants. When the supply of available potassium in the soil is exhausted, potassium will move from the old leaves to the young

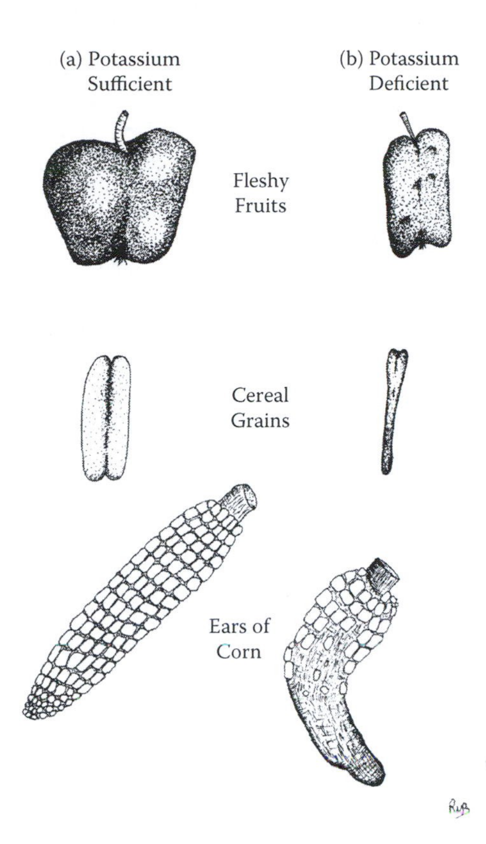

FIGURE 3.16 Expression of potassium deficiency on fruits and seeds, showing malformed fruits (apple), unfilled grains (wheat), and unfilled ears (corn).

ones. Deficiency symptoms appear on the old leaves following the depletion of potassium from these organs. The symptoms appear first as white or brown dots near the tips and along the margins of leaves (Figure 3.17). As the deficiency becomes more severe, the tips and margins of leaves appear as if they were burned or scorched.

If these symptoms are detected early in the growth of a crop, corrections by fertilization are possible, but even then yields cannot be restored to their full potential. Fertilization of crops based on the appearance of deficiency symptoms is not a good practice. Generally, when deficiency symptoms appear, irreparable damage to plant health and crop productivity has occurred.

Small and misshapen fruits and unfilled seeds, described in the preceding section, are indications of potassium deficiency during crop growth and development (Figure 3.16). Lodging of crops is also an indication that potassium was deficient. These symptoms tell the grower that potassium fertilization is needed in the following season.

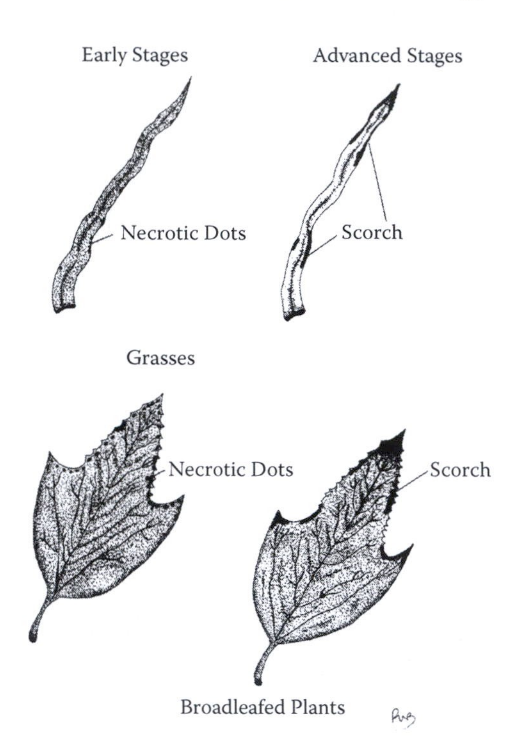

FIGURE 3.17 Foliar symptoms of potassium deficiency on grasses and broadleaf plants, showing the early symptoms of dead spots at leaf margins and advanced marginal scorch

Amounts of Potassium Required by Crops

The amounts of potassium removed by crops is related directly to the productivity of the crops, especially to the amounts of vegetation produced, crop density of planting, and the length of the growing season for maturation of the crops. Potassium enters plants rapidly during their early growth and is accumulated more rapidly than nitrogen and phosphorus. It is recommended that an abundant supply of potassium be available during early growth, particularly since potassium helps to stimulate root growth. However, it is not recommended that all of the potassium be applied at or near planting. Splitting the applications of potassium into two or more during the season gives better efficiency potassium utilization than one large application at the beginning of the season. Placement of potassium fertilizers in bands also increases the efficiency of use of potassium in crop nutrition.

A high demand is for more than 100 lb actual potassium removed per acre (2.5 lb per 1000 sq ft). Generally the crops with high demands for potassium match those with high demands for nitrogen and phosphorus, with corn, tobacco, alfalfa, and celery being notably high in requirements for potassium. A moderate demand for potassium is in the range of 50 to 100 lb actual potassium removed per acre (1.25 to 2.5 lb per 1000 sq ft). Vegetable crops and small grains fall in this category. Less than 50 lb per acre (1.25 lb per 1000 sq ft) is a low demand. Orchard crops have low demands for potassium.

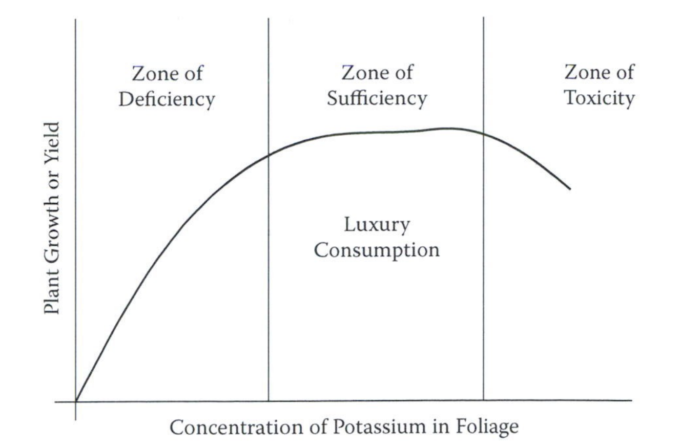

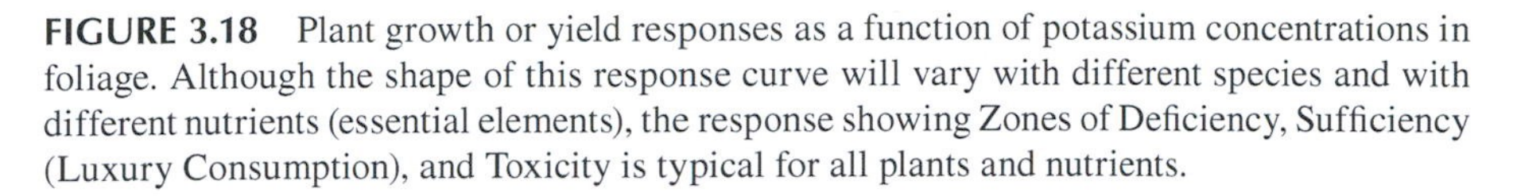
FIGURE 3.18 Plant growth or yield responses as a function of potassium concentrations in foliage. Although the shape of this response curve will vary with different species and with different nutrients (essential elements), the response showing Zones of Deficiency, Sufficiency (Luxury Consumption), and Toxicity is typical for all plants and nutrients.

The vegetation of plants contains more potassium than fruits and seeds. At maturity, crops will have about two-thirds of the potassium remaining in the vegetation, with about one-third being transported to the fruits. Hulls, husks, and rinds have more potassium than the pulp and seeds of fruits. Crops that are fertilized heavily with nitrogen, generally have high requirements for potassium.

Concentrations of potassium in leaves range from 1% to 5% or higher on a dry weight basis. Tobacco plants may have more than 6% K in leaves and even higher values in stems. Sufficiency values in leaves are 1.5% to 3% K for most crops. Values below 1.5% K are considered to be deficient levels. Plants absorb potassium readily, and it may accumulate far in excess of the high range of sufficiency. Excessive accumulation of potassium is called *luxury consumption* (Figure 3.18). Luxury consumption of potassium does no harm to a crop but is wasteful of potassium if the portions of plants with luxury consumption are removed from the sites on which plants are grown. Overfertilization with potassium may suppress calcium and magnesium absorption and lead to deficiencies of these elements in plants, particularly if these elements are in low supply in the soil.

Potassium-Containing Fertilizers

The concentrations of potassium in fertilizers ranges from about 2% to 60% expressed as oxide (K_2O) (Table 3.9). Plant residues, vegetative portions and rinds, hulls, and husks, are good sources of potassium, and all of the potassium in these materials is available, because it is water-soluble. Farm manures have less potassium, about 2% K_2O, than the feed that the livestock received. Much of the potassium is eliminated with the urine of large animals and frequently is lost. Composts are also much lower in potassium than the original plant material, about 1% K_2O, because much of the potassium is lost by leaching during composting. Since

TABLE 3.9
Potassium Concentrations in Fertilizers in Some Common Fertilizers

Source of Potassium	Concentration of Potassium (% dry wt)*		Available K (% of Total)
	K_2O	Actual K**	
Organic			
Plant residues			
Vegetative	4	3.3	100
Hulls and rinds	2	1.7	100
Seeds	1.5	1.2	100
Seaweed (kelp)	5	4.2	100
Manures			
Dehydrated	2	1.7	100
Fresh (wet wt basis)	0.6	0.5	100
Composted	1	0.8	100
Wood ashes	10	8.3	100
Greensand	7	5.8	nil
Granite dust	5	4.2	nil
Chemical			
Potassium chloride (muriate of potash)	60	50	100
Potassium sulfate*** (sulfate of potash)	48	40	100
Potassium nitrate (nitrate of potash)	44	37	100
Potassium magnesium sulfate*** (sulfate of potash magnesia)	22	18	100

* The values reported for the organic fertilizers in this table are typical mean values.
** $0.83 \times K_2O$.
*** Potassium sulfate and sulfate of potash magnesia are certified as organic fertilizers, if the fertilizers are obtained from mined sources.

manures and composts are of plant origin, their potassium is water-soluble and is fully available to plants. However, because of the low concentrations of potassium in manures and composts, large applications of these materials are needed to provide the amounts of potassium required for crop production. Tobacco stems are rarely available, and they should be used with caution because of the tobacco mosaic virus, which is likely to infect tobacco and which is a disease of tomato and other solonaceous (family Solonaceae) crops.

Potassium in granite dust is in the feldspars and micas and is virtually unavailable to plants. Granite dust is a by-product of the quarry industry. Greensand is a naturally occurring mineral, potassium glauconite, occurring in coastal Delaware and New Jersey and other regions of the country. The potassium in greensand is very slowly available, being only slightly more soluble than that in granite dust.

Enormous amounts of granite dust or greensand must be used to supply potassium to a crop, and even then their value in plant nutrition may be small. Some recommendations suggest uses of 5 to 10 tons of these materials per acre per year. Costs of applying granite dust, greensand, and other rock dusts in the amounts that would be needed for potassium fertilization of crops in potassium-deficient land could exceed the value of the crop land.

The chemical fertilizers, except for potassium nitrate, are of natural origin. Some natural deposits of potassium nitrate may occur, but these are not mined for their fertilizer. Potassium chloride (muriate of potash; silvite) is mined and purified. Because of its high concentrations of potassium and chloride and high solubility, it is not considered to be an organic fertilizer. Mine-run potassium chloride is marketed, occasionally, under the name of *manure salts*, which may have 18% to 22% K_2O. Potassium sulfate (arcanite) and potassium magnesium sulfate (langbeinite) are also naturally occurring ores, although commercial sources may be manufactured from potassium chloride and other sources. Potassium magnesium sulfate is sold by several vendors of organic fertilizers, and certifying organizations allow use of the mined material. Potassium sulfate and sulfate of potash magnesia of mined origin are permitted as organic fertilizers by certifying organizations. Defining potassium magnesium sulfate or potassium sulfate as an organic fertilizer is rather nontraditional in the sense that they are no more naturally occurring than potassium chloride and are water-soluble.

Potassium in Soils

The amounts of potassium in soils are high in comparison to those of nitrogen and phosphorus (Tables 3.4, 3.8, 3.10). Most of the potassium in soils, 90% to 98%, is held in *primary minerals*—feldspars and micas (Figure 3.19). This potassium is virtually unavailable to plants until it is dissolved by weathering. It is virtually the same kind of material as granite dust. The potassium in the primary minerals of soils makes a more substantial contribution to plant nutrition than fertilization with granite dust simply because of the very large amounts of potassium in the primary

TABLE 3.10
Amounts of Potassium in Soils of Various Textural Classes

Soil Texture	Potassium Content (lb in Top 6 Inches per Acre)
Sand	3,000
Sandy loam	20,000
Loam	30,000
Silt loam	40,000
Clay loam	50,000
Clay	70,000

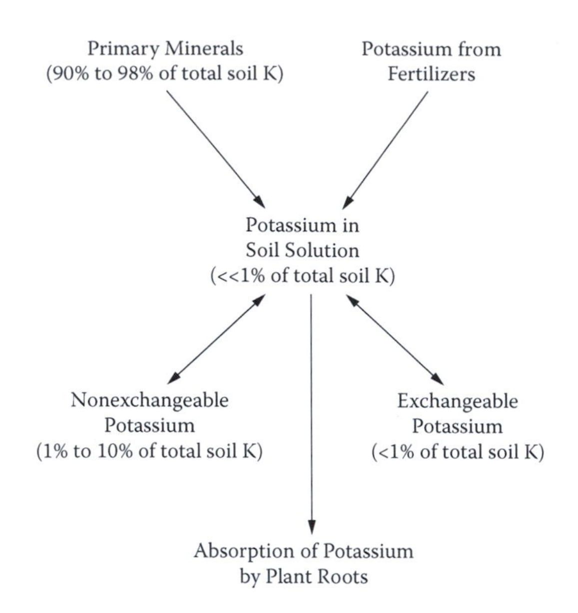

FIGURE 3.19 Equilibrium relationships between forms of potassium in soil.

minerals of soil in comparison to that added with single or multiple fertilizations with granite dust. A sandy soil has 60 to 200 tons, and a clay has about 700 tons of primary minerals in the top 6 inches of an acre of land.

The clays of soil hold another major fraction of potassium. Micaceous clay minerals are made up of sheets, like those of horticultural vermiculite except much smaller. Potassium can be trapped in these plates (Figure 3.20). This potassium is termed *nonexchangeable* or *fixed*. From 1% to 10% of the total soil potassium may be in this fraction. This potassium is not released readily but has substantial availability to plants. The nonexchangeable potassium may contribute as much as 75% of the potassium nutrition of a crop during the growing season. The strength by which the potassium is trapped depends on the kinds of clays in the soil.

Clays in soils are colloidal particles with negative charges. They can attract and hold positively charged ions, *cations*. The ability to hold these ions is called *cation exchange capacity*. Organic matter in soil also is negatively charged. Humus is colloidal and has a high cation exchange capacity, even higher than that of most clays. Clays and organic matter hold a portion of soil potassium that is called *exchangeable* potassium (Figure 3.19). Silts and sands have very low cation exchange capacities in relation to clays and humus. The exchangeable potassium may constitute 1% or 2% of the total soil potassium. All of it is readily available to plants, as the exchangeable potassium can enter readily into soil solution. Holding of potassium to the exchange sites of colloids protects it against leaching.

Plants get their nutrition from the potassium in soil solution. This fraction is much less than 1% of the total soil potassium. The relationships among the various components of potassium in soil are shown in Figure 3.19.

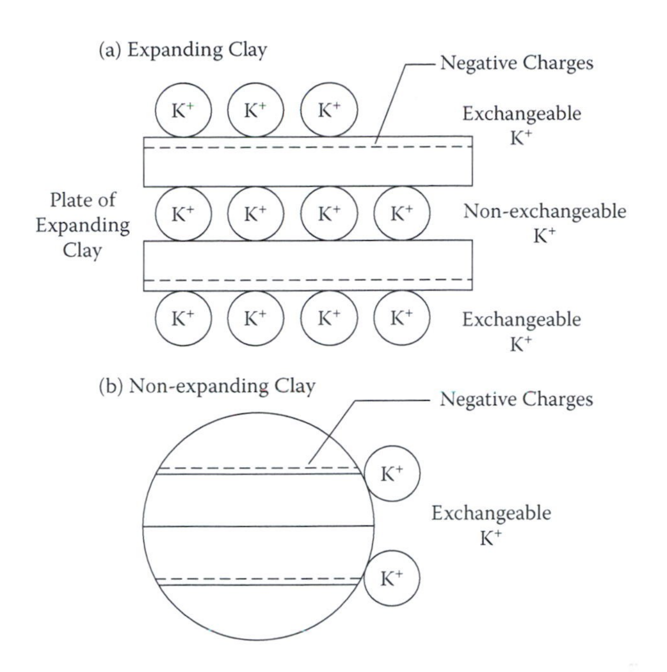

FIGURE 3.20 Exchangeable and nonexchangeable potassium ions (K+) held to or within (a) expanding lattice clay, such as various soil micas, and (b) nonexpanding clay, such as kaolinite.

Problems in Maintaining Fertile Levels of Potassium in Soil

Reliance on Total Potassium Content of Soil

Once, even into the 1940s, it was believed that soils did not require potassium fertilization because of the high total contents of potassium in soils. People who held this belief were unaware of the chemistry of potassium in soil. They did not realize that most of the potassium in soils is held in the primary minerals and as such is unavailable to plants. The fertility of a soil with respect to potassium depends on the amounts of potassium that are held in the exchangeable and nonexchangeable forms and to a much lesser extent on the amount in the primary minerals. The soluble fraction is important only in fertilized soils. In some soils, the exchangeable and nonexchangeable potassium are soon exhausted and not replenished from the reserves in the primary minerals. These soils are said to have a low potassium-supplying power. Several years may be required before the weathering of the primary minerals is sufficient to restore available potassium to a level to support production of a good crop after exhaustion of the exchangeable and nonexchangeable fractions. The length of time for restoration to occur is related somewhat to the total potassium in the soil. Generally, the more total potassium in the soil, the more that is available for weathering, and the faster the restoration. Restoration will be much slower in a sandy soil than in a clayey soil. However, the more intensive the crop production, the faster available potassium will be depleted from the available sources.

Fertilization of crops with nitrogen and phosphorus enhances crop yields and accelerates the rate of depletion of potassium from soils. Squanto, the American Indian whom some people call our first extension agronomist, taught the pilgrims to fertilize corn by placing fish in the hills. Yields of corn were raised, and in a few years, the available potassium was depleted from the soils of Massachusetts. These soils had a poor potassium-supplying power, and the productivity of soils was lost because of depletion of available potassium. In Squanto's time (about 1630), potassium had not been discovered as an element (discovered about 1807), and nothing was known about its requirements by plants (essentiality demonstrated about 1860). However, if the pilgrims had used wood ashes or seaweed in their fertilization program, productivity of fields could have been maintained better than with the fish alone.

After World War II, increased manufacture of nitrogen fertilizers occurred with the diversion of military industries to civilian production. Applications of nitrogen fertilizers greatly increased. Phosphorus fertilization increased also, but because people believed, based on its total analysis, that the soil had plenty of potassium, potassium fertilization did not increase in accordance with the use of nitrogen and phosphorus fertilizers. In many soils of the United States, the depletion of available potassium by enhanced crop productivity soon exceeded the capacity of soil to provide potassium, and the need for fertilization was recognized by scientists and growers.

The concentration of potassium often increases with depth into the soil. More unweathered primary minerals occur deeper in the soil than in the topsoil. Deep plowing turns up these minerals and enriches the topsoil with potassium. This action does not increase the fertility of the soil, because these unweathered minerals have very low capacities to supply potassium to the available pool. In fact, erosion, which removes top layers of soil, may lead one to believe that the potassium fertility of soil has not been depleted by erosion. Removal of the top layers exposes layers with higher total potassium concentrations than the uneroded soil, but in this case, the higher potassium is due to higher relative amounts of unweathered primary minerals, which have very low capacity to supply potassium to a crop. In fact, the potassium-supplying power of a soil is diminished by erosion because of the loss of the fine particles, which are more subject to weathering and more likely to contain nonexchangeable or exchangeable potassium than soil particles from deep zones.

Fixation of Potassium in Soil

Fixed potassium is synonymous with nonexchangeable potassium (Figure 3.20). Fixed potassium is that which is trapped in the plates of clay. In general, the higher the clay content, the greater the fixation. Clays differ in their capacities to fix potassium. Kaolinite, which does not have a lattice or plate structure, does not fix potassium, although it will hold potassium ions in its relatively limited exchange sites. The plate-like clays also differ in their capacities to hold potassium in fixed sites. These differences are related to the structure of the clays in relation to the size of potassium ions.

Potassium that is applied from fertilizers may be trapped in the lattice of clays. This potassium is temporarily unavailable to plants and may create shortages for plant nutrition in soils in which the potassium is fixed strongly. However, fixation by clays holds potassium against leaching so that in clay soils, leaching is virtually nil. In sands, which have limited capacity to hold K^+, as much as 30% of fertilizer potassium may be lost by leaching. Fixed potassium has an important role in nutrition of crops. As much as 75% of the potassium nutrition of a crop may be obtained from fixed potassium as potassium is released from the fixed sites during the entire growing season.

Leaching of Potassium from Soils

All of the potassium in soil solution is subject to leaching. The amount of leaching varies with soil texture. In coarse-textured soils (sands, loamy sands, sandy loams), water moves rapidly through the large pore spaces, and potassium in solution is transported rapidly downward. Coarse-textured soils have weak capacities to hold potassium in exchange and fixed sites; hence, the soil has little capacity to hold potassium against the forces of leaching. Yearly losses of potassium may reach or exceed 100 lb per acre in uncropped sandy soils, leading to rapid depletion of native soil potassium or of potassium added with fertilizers. Fine-textured soils have small pores through which water can move only slowly. This slow movement restricts losses from leaching. Slow percolation of water downward also keeps it in contact with clays and organic matter, which can hold potassium by exchange and by fixation. Losses of potassium from uncropped silty or clayey soils may be only 1 lb per acre per year.

Luxury Consumption of Potassium

If potassium is in an abundant supply in soil, plants will absorb potassium beyond the amounts that they need. Plants may absorb two to four times the amount of potassium that they need for their metabolic requirements. In soils with high capacities to supply potassium, luxury consumption is unavoidable. Most often, luxury consumption results from overfertilization. The principal problem with luxury consumption is that it is wasteful of potassium. If the vegetation of a crop is sold off the farm, the grower is wasting money by overfertilizing with potassium.

Management of Potassium Fertilization

Because of fixation, leaching, and luxury consumption of potassium, frequent light applications of potassium during the growing season are better than one large application at planting. Making several applications is labor intensive and requires more use of equipment than one application. Growers should balance these costs with those of the fertilizer and can consider only one application if costs of labor and equipment exceed the cost of potassium fertilizer. As noted earlier, economies of costs and efficiency of nitrogen use are achieved with several applications of nitrogen during the growing season relative to one massive application at or before planting. In contrast to applications of potassium and nitrogen, all of the phosphorus fertilization of a crop should be applied at planting to ensure that the young plants get an adequate supply of phosphorus.

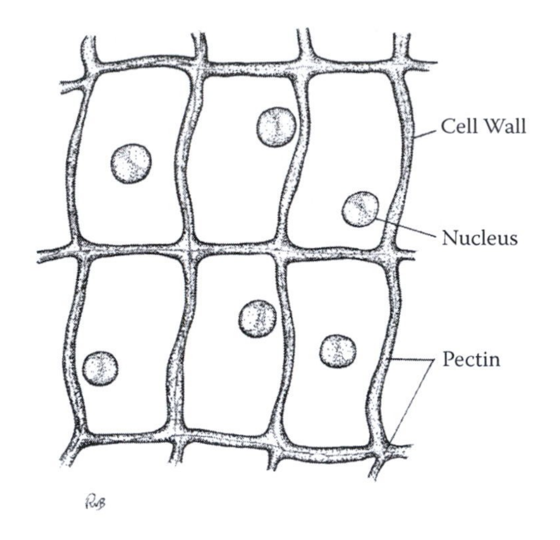

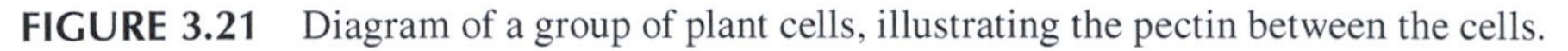
FIGURE 3.21 Diagram of a group of plant cells, illustrating the pectin between the cells.

CALCIUM

Function

Experiments (Sachs, Knop) with solution culture during the 1860s demonstrated the essentiality of calcium. Much of the calcium in plants is in pectin, which is a polysaccharide (complex sugar) located between cell walls (Figure 3.21). In a sense, pectin is a cementing agent between cells. Growth of cells by expansion requires synthesis of pectin, which surrounds the growing cells. Without adequate synthesis of pectin, expanding cells may tear away from others causing some cells to die. Calcium also has roles in cell division. Calcium is needed for nuclear division (mitosis) to form two nuclei and for the formation of a plate that divides one cell into two. Calcium-deficient plants may have cells with one or two nuclei or with no nuclei. Division of chromosomes may be abnormal. Some of these cells may die.

Calcium has functions in transport of materials in plants. It regulates permeability of cellular membranes. Calcium-deficient cells may become leaky and lose their selectivity in ion accumulation. Calcium may have a role in regulating cell acidity. Calcium serves as an antagonist for other cations. A high concentration of calcium is needed especially if potassium, magnesium, or sodium is high in nutrient solutions or in the soil solution. Calcium is needed for the nodulation of roots of legumes.

Symptoms of Calcium Deficiency in Plants

Knowledge of the symptoms of calcium deficiency is a prerequisite to understanding the effects that calcium has on plant growth and quality. Calcium is an immobile element in plants. An element is immobile in plants if it is not transported from one organ of the shoot to another part of the plant via the phloem. Calcium that enters into tissues during the early development of a plant is not translocated to young tissues or fruits even if the calcium supply is depleted from the soil.

Nitrogen, phosphorus, and potassium are mobile nutrients and are transported from old leaves to young leaves, flowers, or fruits as a natural process in plant development and especially if these nutrients are depleted from the soil. Deficiency symptoms of these elements appear first on the old leaves.

Symptoms of calcium deficiency begin at the growing points and the surrounding young leaves. The symptoms appear first as dead spots across the blades of the leaves. The growing points and the young leaves produce die back from the tips downward. Symptoms will progress down the plant until the fully expanded leaves are reached. If leaves are not expanding, their needs for additional calcium are not as great as those of young leaves; hence, they may not show the symptoms. The young leaves of corn and other grasses may fail to unroll, appearing as if they are stuck together. The oldest leaves on plants may have adequate calcium and show no signs of stress, but they cannot donate calcium to the leaves that need it, due to the immobility of calcium.

The immobility of calcium is caused by its being deposited in insoluble compounds in the protoplasm. It may be precipitated as calcium phosphates or oxalate. The calcium in pectin between the cell walls also is immobile. Foliar symptoms such as those described above are rare in field-grown crops, but many nutritional disorders related to the immobility of calcium in plants commonly appear in fruit and vegetable crops and are discussed in the section of effects of calcium on plant growth and quality.

Effects of Calcium on Plant Growth and Quality

Calcium deficiency occurs most commonly in acid, sandy, highly leached soils in which calcium has been depleted to low levels by environmental processes. Calcium deficiency also occurs in dry soils. In this case, water is insufficiently available for dissolution of calcium and for calcium absorption and transport by plants. Calcium is distributed within plants with the flow of water. If sufficient water is not reaching a tissue or organ, calcium deficiency can develop. Calcium deficiency can occur also in soils that are highly fertilized with potassium or ammonium fertilizers. Potassium and ammonium competitively suppress calcium absorption by plants, and if calcium concentrations in the soil are marginal, deficiencies may occur. Deficiencies of calcium frequently cause disorders of several horticultural and agronomic crops and render the products unmarketable or undesirable for consumption.

Growing Point Disorders

Internal tipburn of cabbage, lettuce, and brussels sprouts is caused by calcium deficiency (Figure 3.22). The outer leaves of these vegetables appear normal because they are the oldest leaves of the heads. When the head is cut longitudinally through the core, browning of the tips of the young leaves is manifested. Internal tipburn greatly lowers the value of these crops. Tipburn of looseleaf lettuce is an expression of calcium deficiency. In development of this disorder, the young leaves do not receive adequate calcium from the soil. The soil may be deficient of calcium in this case, or calcium may fail to reach the young leaves as it is directed to the old, larger leaves instead. Blackheart of celery is a similar disorder to internal tipburn. In this case, the prized hearts of celery are blackened by death of the growing points and of the young leaves that constitute the hearts.

Fruiting Problems

Blossom-end rot is an expression of calcium deficiency of tomato, eggplant, pepper, and watermelon (Figure 3.22). The blossom end is the portion of the fruit opposite point of attachment of the fruit to the stem. Blossom-end rot is tough, leathery, dead tissue. Secondary infections with decay organisms usually do not occur with blossom-end rot, but these fruits have no market value and may ripen at very small sizes. Unblemished portions of the fruits usually have a bad flavor.

Blossom-end rot sometimes can be prevented by maintaining well-irrigated conditions during the fruiting of crops. Water is needed for calcium to be in solution to be absorbed by plants. Calcium does not move into the plant with the flow of water, but calcium moves up plants with the flow of water, and in dry soils, insufficient calcium enters plants and is not distributed to the fruits to prevent development of blossom-end rot. Fruits have low rates of transpiration so that the movement of water and calcium into fruits is limited, especially in dry soils. Also, in dry soils, potassium and other cations from salts more soluble than calcium compounds will be taken up preferentially to calcium, thereby suppressing calcium absorption by plants.

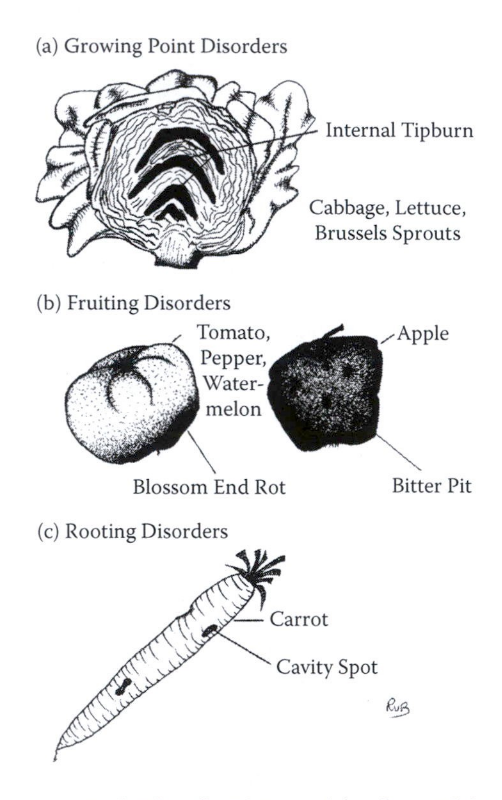

FIGURE 3.22 Some crop production disorders resulting from calcium deficiency.

Blossom-end rot appears often in plants grown in peat-based media that are used in greenhouse production. These media are deficient in calcium, and attention must be given to supplying calcium with fertilization. Calcium nitrate (not organic) is sometimes applied to soils or media in which plants are showing blossom-end rot. Fertilization will prevent further development of blossom-end rot on young unblemished fruits, but fruits that show the disorder will not recover. Liming of soils or media at planting of crops helps to prevent this disorder.

Bitter pit of apple is a nonpathogenic disease from calcium deficiency (Figure 3.22). In this case, small, skin-deep pits with a bitter taste develop in the peels of apples. Other disorders that occur during storage and which greatly shorten the storage life of apples are caused by inadequate calcium in the fruit. Internal breakdown and watercore reportedly are caused by insufficient calcium in fruits. These disorders often are prevented by dipping fruits in slurries of calcium chloride to increase the concentration of calcium in the fruits.

Peanuts require calcium for fruiting. The peanut flowers are above the ground. After pollination, the flowers develop an organ called the *peg* or *gynophore*. The peg burrows into the ground so that peanuts form in the soil. The peg will not grow into soil in which calcium is deficient, and as a result, no peanuts are produced.

Root Problems

Calcium-deficient branch roots are stunted and stubby. Cavity spot of carrot (Figure 3.22) is a disorder that may appear on otherwise normally developed roots. The disorder appears as blackened sunken lesions on carrot or parsnip roots. It is caused by imbalances from fertilization with potassium or with ammonium relative to calcium supply in the soil. Heavy applications of fresh farm manure are suspected to promote formation of cavity spot in soils with marginally low levels of calcium. The potassium or ammonium in the manures may suppress calcium absorption by the carrot to the point of deficiency.

Effects of High Concentrations of Calcium in Soils

High concentrations of calcium have little direct effects on plant growth. Few if any toxicity symptoms are associated with excesses of calcium in soil or in the soil solution. Problems are associated usually with the accompanying anion. High levels of calcium in association with chloride or nitrate may present temporary problems until these ions are leached from the soil. These problems may be associated with salinity from the high concentrations of ions in solution. Calcium in association with sulfate rarely presents problems, because calcium sulfate is sparingly soluble. Sulfate also has low toxicity. Soils with free calcium carbonate may have alkaline pHs that limit availabilities of minor elements and phosphorus. Possibly, but rarely, high concentrations of soluble calcium may depress potassium or magnesium absorption by plants in soils in which potassium or magnesium supplies are marginal for plant nutrition.

Normal ranges for calcium in the top 6 inches of an acre of unlimed soil are 1000 to 8000 lb in humid regions. Fine-textured soils have more calcium than coarse-textured soils. Although calcium is held tightly and is the dominant cation on soil particles, it leaches from soils in humid areas so that it is the most abundant cation in tile drainage waters and streams. Surface layers of soil will have less calcium than lower horizons. In arid regions, calcium in the topsoil may be up to 20,000 lb per acre mainly as calcium sulfate or calcium carbonate or bicarbonate.

The cation exchange capacity of soils should be dominated by calcium for good fertility. About 80% of the exchange sites being saturated by calcium is a good value. About 15% of the exchange capacity should be held by magnesium, and about 5% of the exchange capacity should be held by potassium.

Calcium Removal by Plants

Plants remove from 30 to 50 lb of calcium per acre per year. The higher the yield of crop vegetation or fruits, the more calcium is removed.

Calcium Concentrations in Fertilizers

The organic grower generally does not make a deliberate effort to apply calcium fertilizers. Generally, the calcium requirements of plants are met through applications of materials to supply nitrogen and phosphorus and to regulate soil acidity. However, conventional growers who use concentrated soluble fertilizers for nitrogen, phosphorus, and potassium, particularly if the fertilizers are applied as liquids, may have to take specific action to ensure that soils provide adequate calcium. Some materials that provide calcium are listed in Table 3.11. The calcium contents will vary with the origin, purity, and hydration of these materials.

Finely ground limestones and other limes added to correct soil acidity will provide adequate calcium to crops, providing no other factors such as water supply are

TABLE 3.11
Calcium Concentrations in Limes, Soil Amendments, and Fertilizers

Material	Concentration of Ca (%)	Availability of Ca
Limes		
Agricultural limestone	40	Slow
Dolomite	22	Slow
Quicklime	70	Rapid
Hydrated lime	50	Rapid
Amendments		
Gypsum	30	Moderate
Fertilizers		
Ordinary superphosphate	20	Moderate
Triple superphosphate	16	Moderate
Rock phosphate	33	Very slow
Bonemeal	38	Slow to moderate
Calcium nitrate	24	Soluble
Plant residues (vegetative)	1 to 4	Moderate
Manures	1	Moderate

limiting. Gypsum is added to fine-textured soils to improve their structures. Gypsum is called a *soil amendment* rather than a fertilizer, since its use is to improve soil structure rather than to supply calcium. Clays on which calcium is the dominant cation in the cation-exchange sites are flocculated to form aggregated structures, which improve the structure of fine-textured soils. Gypsum adds calcium to the soil so that structural improvements can be made without raising pH. Gypsum can be a source of calcium for plants growing in soils in which liming is unneeded or undesirable or may be used as a calcium fertilizer regardless of soil acidity.

Crops that are fertilized with bonemeal or superphosphates to supply phosphorus will receive enough calcium. Availability of calcium from rock phosphate is limited by its low solubility, and unless practices are taken to improve the availability of phosphorus from rock phosphate, calcium may be supplied inadequately. Plant vegetative residues, composts, and manures applied in quantities that will meet the nitrogen requirements of crops also will supply adequate calcium to the crops. Fruits and seeds are low in calcium, having only 0.1% calcium or less, and are not suitable to consider as calcium-containing fertilizers because of the large amounts of these materials that would have to be applied to supply sufficient calcium for plant nutrition.

MAGNESIUM

Functions

The essentiality of magnesium was demonstrated with solution culture of plants in the 1860s (Sachs, Knop). Magnesium is a constituent of the green pigment, chlorophyll, of plants. The requirement of magnesium for chlorophyll synthesis is absolute. No other element will substitute for magnesium in this role. About 1% to 3% of the magnesium of plants is in chlorophyll. Although only a small fraction of the magnesium of plants is in the chlorophyll, its function in this role is a major one. Chlorophyll is required for photosynthesis, a process by which plants convert light energy into chemical energy. The chemical form of energy generated by photosynthesis is a compound known as *adenosine triphosphate* or *ATP*. ATP is also a product of chemical transformations of energy during respiration. Magnesium is required for biosynthesis and metabolism of ATP during photosynthesis or respiration. Magnesium is essential for activation of many enzymes, which are organic catalysts in metabolism. Magnesium activates more enzymes in plants than any other nutrient. Many of these enzymes also are involved in phosphorus metabolism. Magnesium through its action in activation of enzymes and biosynthesis of ATP is essential for protein synthesis in plants.

Symptoms of Magnesium Deficiency in Plants

Magnesium deficiency may occur in highly leached, acid, sandy soils from which it has been depleted. Organic soils with marl deposits may be magnesium deficient because of the antagonism of calcium with magnesium in absorption by plants. Soils with low magnesium contents and limed with calcitic limestone may become deficient because of the antagonism between calcium and magnesium. Potassium and ammonium fertilization also can produce magnesium deficiency by cation antagonism. Pastures that are growing in soil in which magnesium is just sufficient for crop growth may have depressed concentrations of magnesium in the forage following liberal applications of potassium fertilizers. In some cases, the forage may not have sufficient magnesium for livestock nutrition, creating a condition known as *grass tetany*.

Magnesium is a mobile element in plants. Symptoms of deficiencies will appear first on the old leaves, as magnesium may be translocated from old leaves to young leaves or to reproductive or storage organs. Deficiencies appear as yellowing of the old leaves. Yellowing occurs initially between the veins with green color remaining along the veins. This pattern of expression of deficiency is called *blotching* or *mottling* (Figure 3.23). The lower leaves of grasses will have alternating yellow and green stripes, with the blades being green along the veins and yellow in between. Some plants (cotton, grapes) may show purpling over the blades. In advanced stages of magnesium deficiency, dying of tissues occurs so that the symptoms are indistinguishable from those of potassium deficiency. If symptoms appear on plants, it is usually too late to take corrective action.

The yellowing of leaves is caused by loss of chlorophyll and protein. Magnesium is an essential constituent of chlorophyll, and the yellowing of leaves in part reflects the shortage of magnesium for chlorophyll synthesis. Much of the disappearance of chlorophyll can be traced to inadequate protein synthesis for which magnesium is required. All chlorophyll is held in a complex with proteins in the chloroplasts. About 10% of the magnesium in plants is in the chloroplasts, where it has functions in photosynthesis and protein synthesis. Since about 3% of the magnesium is in the chlorophyll, the major fraction of the magnesium is functioning in enzyme activation and protein synthesis.

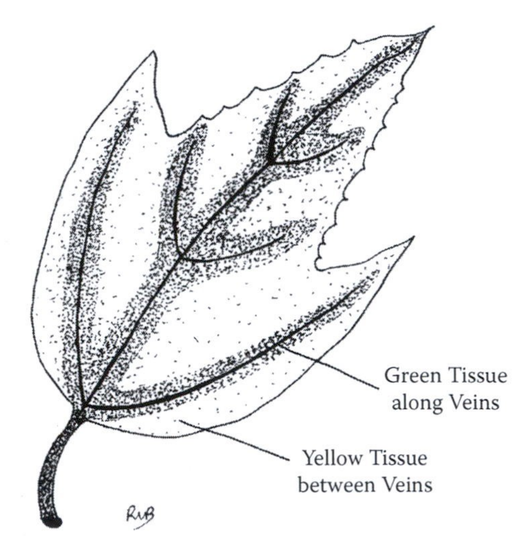

FIGURE 3.23 Magnesium-deficient leaf, showing a mottled appearance of green and yellow or dead tissues.

Effects on Plant Growth and Quality

Magnesium deficiency brings about essentially irreversible losses in crop yields. Vegetative portions with low magnesium contents will have poor color and quality. Forages that are grown in magnesium-deficient soils may have insufficient magnesium for nutrition of farm animals, producing grass tetany of livestock.

Amounts of Magnesium Required by Crops

Crops remove between 15 and 30 lb of magnesium per acre per year. Magnesium concentrations in leaves are about 0.4% on a dry weight basis. Concentrations below 0.15% to 0.25% are too low for adequate nutrition of most plants. Concentrations over 1% may be toxic levels and indicate excessive magnesium in soils. Excesses of magnesium are rare but may arise in soils derived from rocks high in magnesium or from excessive magnesium fertilization. Organic growers should rarely encounter excesses arising from overfertilization, unless they have applied higher than agronomic rates of magnesium fertilizers or have used dolomitic (high magnesium) limestones repeatedly and frequently. Applying calcitic limestone or gypsum can help to correct problems of excessive magnesium in soils.

Most of the magnesium in soils is in the primary minerals. Magnesium contents in soils are about 6000 lb per acre (top 6 inches) in humid regions and about 12,000 in semiarid regions. Magnesium is held as an exchangeable cation to clays.

Concentrations of Magnesium in Fertilizers

The most economic method of providing magnesium to crops is through the application of magnesium-containing limestones (Table 3.12). Limestones vary in magnesium contents depending of the origin of the materials. Pure calcium carbonate is a mineral called *calcite*. Dolomite is a 1:1 mixture of calcite and magnesium carbonate. Most limestones are mixtures of calcite and dolomite. These mixtures are called *intergrades* or more specifically are called *dolomitic limestones*, *magnesium*

TABLE 3.12
Magnesium Contents of Limestones and Fertilizers

Material	Magnesium Concentration (%)
Limestones	
Dolomite	12
Dolomitic limestones	1.3 to 6.5
Average agricultural limestone	4.9
Fertilizers	
Magnesium sulfate (Epsom salts)	10
Potassium magnesium sulfate	11
Talc and soapstone	19

limestones, *high-magnesium limestones*, or other terms. Growers should be aware of the composition of limestones and make their purchases accordingly, selecting limestones with magnesium unless a soil test notes that magnesium is not required. The bag in which limestone is packaged should indicate the magnesium concentration in the lime. Growers should ask for a report of the chemical analysis of limestone bought in bulk

Limestones should be ground finely and applied some time before a crop is planted. Magnesium-rich limestones are somewhat harder and less soluble than agricultural limestone, but the differences among the limestones are so small as to not be important agronomically.

Magnesium sulfate, also called *Epsom salts* ($MgSO_4 \cdot 7H_2O$) is water-soluble (Table 3.12). It may be sprayed on plants to correct deficiencies, as magnesium is a mobile element. About one-third ounce of Epsom salts can be dissolved in 5 gallons of water and sprayed on about 100 sq ft of surface (measured on the ground) or to wet the foliage of upright plants. Potassium magnesium sulfate is also water-soluble, but it is normally used as a soil-applied fertilizer. By applying liberal definitions, some organizations recognize magnesium sulfate and potassium magnesium sulfate as organic materials, as these fertilizers are naturally occurring. Soapstone and talc are used rarely as fertilizers, but they are naturally occurring magnesium silicates. Soapstone and talc are soft, and if ground finely and mixed well in the soil, they will release magnesium sufficiently to nourish a growing crop.

SULFUR

FUNCTIONS

The essentiality of sulfur was demonstrated by experiments with solution culture of plants during the 1860s (Sachs, Knop). Sulfur is a constituent of two amino acids (cysteine and methionine) that are required for protein synthesis. In addition to its role in the primary structure of proteins, sulfur is important in bonds that form and maintain the structural integrity of enzymes and other proteins. Sulfur is a constituent of several vitamins and other cofactors of enzymatic reactions in plants (thiamin, biotin, lipoic acid, coenzyme A). Sulfur is a part of several compounds that give the characteristic flavors and odors of mustards (including cabbage, broccoli, and other brassicas), onions, and garlic.

OCCURRENCES OF SULFUR DEFICIENCY IN PLANTS

Sulfur deficiencies have not occurred frequently in crops. Organically grown crops seldom show sulfur deficiencies because of the widespread use of sulfur-containing plant residues, composts, and farm manures in organic agriculture. Irrigation water in semiarid regions and rain water in humid regions often contain enough sulfur to satisfy plant nutrition. The amount of sulfur in precipitation varies by location, but annually around 15 lb of sulfur are rained down in the northern Midwest and in the northeastern regions of the United States. This amount of rained-down sulfur is enough to meet half or more of the sulfur requirement of many crops. Other regions of the country receive less than this amount. Much of the sulfur in the atmosphere comes from the burning of coal in generation of electric power, with some coming from the burning of oil and gas at power plants and from the emissions of automobiles. A small amount of sulfur may enter the atmosphere from decomposing organic matter in swamps. Crops grown in sites that are remote from industrial areas or that are out of the wind flow from industrial areas may receive only a small fraction of their sulfur from precipitation and rely entirely on the soil for their sulfur. Under such conditions and with enhanced yields of crops, sulfur is being depleted from much crop land, particularly if the crops are fertilized with modern, high-analysis, chemical fertilizers, which are inherently low in sulfur. Modern concentrated fertilizers for nitrogen, phosphorus, and potassium are made commonly from urea, concentrated superphosphate, and potassium chloride and, hence, are essentially void of sulfur. As crop yields increase and soils are mined (crop removal without addition of nutrients) for sulfur, deficiencies are becoming more common than in the past.

Some soils are naturally low in sulfur. Sandy soils, low in organic matter, are likely to be sulfur-deficient for crop production. Some soils in the northwest, southeast, and upper Great Plains are formed from rocks that were low in sulfur. Reports of sulfur deficiency have been historically reported for crops in these soils.

Fertilization of crops with materials applied specifically for their sulfur contents is becoming a practice in much of commercial agriculture due to the fact that modern concentrated fertilizers are essentially void of sulfur. For conventional agriculture, a large number of materials contain sulfur. Ordinary superphosphate (12% S), potassium sulfate (18% S), and ammonium sulfate (23% S) are well-known sulfur-containing materials. However, these materials are not used extensively in modern concentrated fertilizers. High-analysis fertilizers contain very little sulfur, because they are formulated from salts that do not contain sulfur, mainly with the purpose of limiting the costs of shipping of the primary nutrients, nitrogen, phosphorus, and potassium. In regions of intensive agriculture that rely on commercial high-analysis fertilizers, concerns have arisen about exhausting sulfur reserves from the soil due to the fact that little or no sulfur is added with these fertilizers.

Symptoms of Sulfur Deficiency in Plants

Sulfur-deficient plants have a light green color that somewhat resembles that of nitrogen-deficient plants. However, sulfur deficiency generally appears first on the young leaves of plants, whereas nitrogen deficiency appears first on the old leaves. The similarities in symptoms of sulfur deficiency and nitrogen deficiency are due to the requirements of sulfur and nitrogen in protein synthesis. The green pigment, chlorophyll, in plants is in a complex with protein in the chloroplasts, and if protein synthesis is diminished, chlorophyll contents in leaves diminish. Sulfur is only moderately mobile in plants, unlike the high mobility of nitrogen. Sulfur deficiency is not transported rapidly from old leaves, and the symptoms commonly appear on the young leaves at the top of plants or tips of branches. With corn and other grasses, the upper leaves may be striped with green and yellow. With advanced stages of deficiency, yellowing may appear over the entire plant of any crop.

Sulfur-deficient herbaceous plants are spindly and stunted. The symptoms may appear before those of yellowing or may be the only symptoms and may go undetected without a standard of comparison. Sulfur deficiency is noted most often on young plants. The symptoms may disappear as organic matter decomposes and as roots permeate the soil. Although the symptoms of deficiency may disappear with age of plants, crop yield potential is not restored, because the detrimental effects of deficiency remain for the life of the plants. As with the occurrence of any nutrient deficiency, sulfur deficiency produces irreparable damage.

Effects of Sulfur on Plant Growth and Quality

Yields of sulfur-deficient plants will be diminished relative to those of well-nourished plants. The stunted and spindly plants are unable to produce high yields. With forages, stands may be thin, and growth may be weak. Sulfur-deficient crops are delayed in flowering, fruiting, or maturation. With perennial crops, deficiencies may appear in the second or third year of cropping. Sulfur deficiency reduces the nodulation and capacity of legumes to fix nitrogen; therefore, sulfur-deficient legumes also may be nitrogen deficient.

Sulfur Removal by Crops

Crops require about 15 to 30 or more pounds of sulfur per acre per year. In general, the sulfur and phosphorus requirements are comparable among crops. Alfalfa, clover, soybeans, corn, sorghum, sugar beets, turnips, cabbage, and onions are examples of crops with high requirements (30 lb/acre or more) for sulfur. The legumes have relatively high concentrations of sulfur and are relatively high yielding. Legumes have about equal amounts of sulfur and phosphorus. The high sulfur requirements of the brassicas and onions are based on their high concentrations of sulfur-containing compounds. These crops may accumulate two or three times as much sulfur as phosphorus. Corn and sorghum have low concentrations of sulfur, about equal to the phosphorus, in their tissues, but they are high-yielding crops so that the amount of sulfur removed in biomass is large. Small grains, orchard, and low-yielding vegetables may remove only 15 lb sulfur per acre per year.

The concentration of sulfur in healthy leaves ranges from 0.2% to 0.5% of the dry weight. Although concentrations range widely with age and species, values below 0.15% are likely indications of deficiencies of sulfur in most plants.

Concentrations of Sulfur in Fertilizers

Organically grown plants obtain most of their sulfur from soil organic matter and from rain and irrigation waters. Elemental sulfur is an organic material. It could be applied to sulfur-deficient soils to correct deficiencies. Sulfur is used as a fungicide and in such use provides nourishment to crops. Sulfur must be oxidized to sulfate by microbial action in the soil before it is of value as a nutrient. Agricultural gypsum is an organic source of sulfur. Sulfur concentrations in some common materials are given in Table 3.13.

TABLE 3.13
Sulfur-Containing Fertilizers and Other Carriers of Sulfur

Material	Concentration of S (%)	Relative Availability
	Organic	
Elemental sulfur	100	Rapid
Agricultural gypsum	17	Rapid
Plants, composts, manures	0.4	Slow
	Chemical	
Magnesium sulfate (Epsom salts)	14	Water-soluble
Potassium sulfate*	17	Water-soluble
Potassium magnesium sulfate*	22	Water-soluble
Ammonium sulfate	24	Water-soluble
Ordinary superphosphate	12	Rapid

* Accepted as organic fertilizer by some certifying organizations if the origins are from mines.

The concentrations of sulfur in the organic materials—elemental sulfur, agricultural gypsum, and plant residues, composts, and manures—will vary with their origins and purities.

Sulfur in Soils

Most of the sulfur in humid regions is in the organic matter in the surface zones of soil. Typical contents of sulfur in the top 6 inches of an acre of mineral soil in humid regions are about 800 lb. In humid regions, maintenance of an abundant level of organic matter in soils is essential to maintenance of a supply of sulfur to crops. In semiarid regions, sulfur contents may be about 1600 lb per acre and are concentrated also in the surface zones, and much of the sulfur may be present in calcium sulfate (gypsum), which is somewhat sparingly soluble and not readily leached.

Sulfur in organic matter is not immediately available to plants and must be mineralized by soil microorganisms. Sulfur transformations in soil are much like those of nitrogen. Mineralization of sulfur is much like ammonification in the mineralization of nitrogen. Inorganic sulfur can be immobilized by its assimilation by soil microorganisms in soil to which carbonaceous plant residues (mature plants, dead leaves, straw, sawdust) are added. Immobilized sulfur can become available later as the C:S ratio is narrowed in organic matter, just as nitrogen becomes available as the C:N ratio narrows.

Sulfate is not adsorbed to clays and organic matter; hence, it is subject to leaching. Due to the sometimes limited water solubility of sulfate-containing compounds, it is not leached as readily as nitrate but is more subject to leaching than phosphorus. Sulfur is not fixed in soils, unlike phosphorus, which is fixed strongly by iron and aluminum. Thus, sulfur in soils does not become unavailable by fixation. If sulfur is present in organic matter or in gypsum or other sulfates, it will be available to plants when environmental conditions (temperature, moisture) are proper.

IRON

Iron in Soils

Iron is one of the most abundant elements on earth. However, the most important aspect of iron in soils is not its abundance but its limited solubility and availability to plants. Soils may have from 10,000 to 200,000 lb of iron in the top 6 inches of an acre of soil, with a typical amount being 50,000 lb per acre. Most of this iron is present as iron oxides (rust-like compounds), which are difficultly soluble. A few pounds of iron in soil are available for plant nutrition.

The solubility of iron in soils varies with soil aeration, which affects the oxidation state of iron. In well-aerated soils, Fe(III) (Fe^{3+}, ferric iron) dominates. Ferric iron is less soluble in soils than Fe(II) (Fe^{2+}, ferrous iron), which is relatively high in soils that are not well drained. Color of soils due to the presence of iron is used sometimes as an index of the drainage or aeration of a soil. Ferric iron gives a reddish color to well-drained or well-aerated soils. Ferrous iron gives a blue-green color mottled with red in poorly drained or poorly aerated soils. This blue-green discoloration appears predominately in the subsurface zones.

The vast quantities of iron in soil ensure an adequate supply, and iron deficiency usually does not occur in soils as a result of ferric iron dominating in well-drained soils. Deficiencies may occur in alkaline soils in which iron is precipitated or in acid, sandy soils in which iron is depleted by weathering and leaching. Organic soils (generally soils with greater than 20% organic matter in the surface horizon) might be iron deficient due to their low concentrations of iron as a result of leaching during soil formation and from the binding (sequestration or chelation) of iron to the soil organic matter.

The solubility of iron declines rapidly with increasing soil pH. If soil alkalinity rises above pH 7.5, growers may find that their crops are iron-deficient and that they are limited in the kinds of crops that they can grow in these soils. The so-called acid-loving crops—blueberries, rhododendrons, azaleas, and other ericaceous plants (family Ericaceae)—will likely be iron deficient in soils above pH 6.5 to 7.0. Acidification of these soils generally makes iron available to crops. Adding iron salts directly to the soil without acidification of the soil usually is not beneficial, as the added iron will react with the high-pH soil and become unavailable. Foliar sprays of iron sometimes are helpful in correcting deficiencies, but the best way of correcting iron deficiency is through application of organic matter to soil for several years. Plant organic matter (crop residues, farm manures, composts) adds iron to soil and increases the availability of iron in the soil through acidification as the organic matter rots. The organic matter also will help to keep the iron in solution.

Organic soils (peats and mucks or soils with more than 20% organic matter by weight and commonly with 80% or more organic matter) and organic potting media often are iron deficient. In these soils, iron may be held tightly to the organic matter in complexes that are known as *chelates* (derived from a Greek word meaning *claw*).

Chelates are organic compounds with a metal (iron, zinc, copper, for example) bound with varying strengths. The abundance of organic matter in organic soils may bind iron with such tenacity that it is not sufficiently available to support plant growth. Also, many organic soils are highly leached so that iron levels in the soil are naturally low, and chelation to organic matter further reduces the available iron to deficient levels. Plants grown in peat-based potting or seeding media (peat-vermiculite, peat-vermiculite-perlite) often become iron deficient, because of the limited supply of iron in these media and the failure to add iron in fertilizers. Grasses and corn are particularly prone to be iron deficient in these media, although deficiencies may occur with many crops.

Iron that is chelated is protected from reaction with other inorganic constituents in the soil or vice versa. An example of affecting other ions is the protection of phosphate from precipitation by dissolved iron in acid soils. Addition of an abundance of organic matter with phosphate fertilization of soil helps to maintain phosphorus in solution. The protection comes from chelation of dissolved iron and subsequent inhibition of the reaction of iron and phosphates. Chelates of iron with small organic molecules form water-soluble complexes, and in this form, iron is available to plants. Soluble chelates or iron are manufactured commercially to supply iron in foliar sprays and for use in nutrient solutions in hydroponics culture of plants. In hydroponics, the concentration of phosphate far exceeds that of iron, and if the iron was not chelated, it would be precipitated by the phosphates. The transport of iron in plants occurs with iron chelated to citric acid. In this case, the chelation of iron protects it from precipitation in the plants, particularly from precipitation as iron phosphates.

Functions of Iron in Plants

Iron is essential for synthesis of chlorophyll, although it is not a constituent of chlorophyll. Iron-deficient plants have a bleached, white or yellow, appearance. The linkage of iron with chlorophyll synthesis was noted as early as 1844, and the essentiality of iron as a plant nutrient was demonstrated conclusively with experiments with solution culture in the 1860s (Sachs, Knop) using the same procedures used to demonstrate the essentiality of the macronutrients.

Iron is required for photosynthesis. Iron is a constituent of proteins (ferredoxin and cytochromes) that participate in photosynthetic reactions. This role is in addition to that required for the synthesis of chlorophyll. Iron is required for respiration. The conversion of the energy of carbohydrates and other compounds to ATP requires iron. Iron is needed for *nitrogen-fixation* by legumes and other plants that acquire their nitrogen from the air. Iron is a constituent of enzymes called *catalase* and *peroxidases*. These enzymes protect cells against oxidation and might have roles in providing disease resistance.

Effects of Iron on Plant Growth and Quality

Plants absorb a few pounds of iron per acre per year. Iron-deficient plants are chlorotic (yellow or white), especially on the young leaves near the tips. Grasses may show striping of yellow or white and green in the leaves. In advanced cases, the symptoms may advance down all plants until the entire plant shows symptoms. The symptoms may persist for some time before death of the tissues occurs. If iron is supplied before death of the tissues, appearance of the plants may be restored, although productivity may not be restored fully for a given season.

Iron-deficient ornamental species (for example, rhododendron, azalea, camellia, gardenia) are unattractive and unmarketable. Prolonged deficiency leads to die-back of the shoots, and new growth will be weak and chlorotic. Iron-deficient ornamental monocots (grasses, bamboo) will be unattractive or unmarketable. These plants may die back, and regrowth will show iron deficiency more than the original growth. Yields of corn and small grains (for example, barley, oats, rye, wheat, sorghum) will be limited markedly by small and sparse ears and heads. Fruit and nut crops (for example, blueberry, cranberry, citrus, grape, peach, pecan, and walnut) will be low yielding and may be injured severely by chlorosis and die-back.

Acidic, leached, sandy soils of coastal plains may be iron deficient because of their depletion of iron by weathering and leaching. Alkaline soils, particularly those above pH 7.5, are likely to be iron deficient because of the precipitation of iron into sparingly soluble compounds. In organic soils, iron may be deficient because of its chelation to organic matter and low supply due to leaching.

Fertilizers for Iron

Adding inorganic iron salts to soils is rarely a beneficial practice in plant nutrition. The added iron is quickly rendered unavailable by reaction with other soil constituents. Farm manures are good sources of iron. The value of manures in iron nutrition of plants resides not only in the iron added but also in the organic matter that is added simultaneously. Organic acids released by decay of manures help to dissolve soil-borne iron and to keep iron in solution by its chelation. Any decaying plant residues will provide the same benefit as manure in this respect. Applications of farm manures at 10 to 20 tons per acre per year are effective in controlling iron chlorosis in soils of all types—sandy, alkaline, or organic—where deficiencies are likely to occur. Sewage biosolids (sludges) are also sources of iron but are not organic sources. Iron salts often are added to stabilize biosolids at waste water treatment plants and hence biosolids may be quite high in iron. Growers should avoid biosolids from sources where inputs of heavy metals (zinc, lead, cadmium, for example) are substantial. Sewage biosolids are not permitted in organic certification programs because of the suspected contents of metals in the biosolids, in spite of the federal and state regulations that govern the acceptable limits of metals in biosolids destined for use on farm land.

Organic (carbon-containing) iron sources are available as fertilizers. These compounds are chelates of iron. They may be known as *sequestered iron* or as various abbreviated names, EDTA, EDDHA, DTPA, or HEDTA. The specific formulations are made for use on soils of different pHs. Users should consult the dealers for specific uses. The prices of these compounds are many times those of inorganic iron salts, but the chelates are many times more effective than the inorganic salts. The chelates may be applied to soils or media or directly to foliage. Best results often are obtained by direct application to foliage, but even this mode of application has its limitations. It is not certain that foliar applications will enter the plant. If iron does enter the leaves, it is translocated sluggishly in plants, as is shown by the fact that deficiency symptoms appear on the young leaves first. Foliar applications of iron may bring corrections only to the portions that receive the application, as translocation of iron to other areas may be unlikely to occur. Foliar applications of iron are used to correct deficiencies on fruit and nut trees and on woody ornamentals and to green-up turfgrass. Soluble chelates of iron, rather than iron salts, should be used for foliar applications. A wetting agent, soap or detergent, should be used to ensure even spreading of the sprays. Iron toxicity may occur where droplets of the spray accumulate and dry.

Growers should check with their certifying organization to ascertain if soluble iron salts or chelates can be used as a permitted organic practice. Many certifying organizations permit use of salts or chelates of micronutrients if micronutrient deficiencies are documented by visual observation or tissue analysis.

ZINC

Zinc in Soils

Zinc concentrations in soils are much lower than those of iron. Representative total zinc in soil is about 150 lb per acre (top 6 inches) with a range from 20 to 500 lb per acre. Very little of this zinc is available. The availability of zinc in soils follows patterns like those of iron. The availability decreases as pH increases, due to the lessened solubility of zinc. Zinc is held in chelates with organic matter. Most of the available zinc is associated with soil horizons having the highest organic matter contents; otherwise, zinc is fairly evenly distributed in the soil profile. Only trace amounts are held in exchange complexes with clays, although zinc may be precipitated on clay surfaces. Zinc forms sparingly soluble precipitates with phosphates, and high amounts of phosphorus in soils or in plants have been suggested as inducing zinc deficiency in crops. Likewise, high concentrations of phosphorus in plant tissues have been associated with zinc deficiency.

Zinc deficiencies are not widespread in crop production, and responses of increases in yield or growth with zinc fertilization are rare. Tree fruit and nut crops have been reported to be zinc deficient perhaps more frequently than other crops. Zinc deficiencies are reported in sandy soils that are low in available zinc (leached), in organic soils in which zinc is held in a complex with organic matter (chelated), in alkaline soils greater that pH 7.5 (precipitated), and in soils high in phosphates (precipitated). Cold weather restricts zinc absorption by plants, and deficiency symptoms may appear on crops during the cool, early parts of the growing season. Deficiencies are often alleviated as the soil warms.

Functions of Zinc in Plants

The discovery of the essentiality of zinc is credited to Sommer and Lipman (in 1926). In the early 1930s, workers were able to correct little leaf disease of citrus and peach with applications of zinc and were able to demonstrate the same deficiencies when zinc was withheld from plants growing in solution culture. Zinc is a component or is required for activation of several enzymes. These enzymes have roles in many plant processes, including nitrogen metabolism, carbohydrate metabolism, DNA and RNA synthesis, hormone synthesis, and possibly photosynthesis.

Effects of Zinc on Plant Growth and Quality

Plants absorb a few pounds of zinc per acre per year. Deficiency symptoms appear on the young leaves of plants. Zinc is a relative immobile element in plants, and it is not transferred rapidly enough from old tissues to young ones to prevent deficiencies in young leaves once supplies of available zinc are exhausted from soils. Leaf size and twig length are restricted in zinc-deficient fruit and nut trees (hence, the name *little leaf disease*). Die-back occurs in severe deficiency. New growth may be bushy and upright (witches' brooms). Fruit and nut production will be curtailed severely, and in severe cases fruits will be deformed, small, yellow, and dry and woody. Slight deficiencies may not affect fruit or nut quality.

Grasses (corn, sorghum, small grains, forages, and turf) show bands of chlorosis around the midribs of young leaves. Young leaves of broadleaf plants are uniformly chlorotic. Under severe deficiencies, whole plants may become chlorotic. All plants may show shortened internodes, which give the plants a rosetted appearance. This effect may be due to the reported requirement for zinc as a factor in growth regulator (auxin) formation or stability.

Fertilizers for Zinc

Farm manures are a major source of zinc for organic growers. Applications of manures for several years at rates of 10 to 20 tons per acre per year will enrich soils so that they will supply adequate zinc nutrition to crops. Sewage biosolids (sludges) are rich in zinc; however, growers should be aware of the composition of the biosolids with respect to other heavy metals to avoid contamination of their soils with these metals.

Several manufactured fertilizers for zinc are available. The most common of these is zinc sulfate. Chelated sources of zinc (Zn EDTA) are also available commercially. Chelated zinc has a better availability in soils, especially alkaline soils, than zinc sulfate and is a preferred source for foliar applications. Many organic certifying organizations permit the use of zinc fertilizers, other than nitrate or chloride salts, if zinc deficiency is diagnosed in crops.

COPPER

Copper in Soils

Copper concentrations in soils are about 10 to 300 lb per acre (top 6 inches), with a representative value being 100 lb per acre. Copper in soil is adsorbed tightly to organic matter so that very little soil copper is available to plants. High levels of organic matter (greater than 20%) in soil may give rise to copper deficiency. Deficiencies are associated most commonly with peats and mucks, which are organic soils. Copper is released from organic matter as soil pH increases above pH 7.3, but the copper that is released has low availability because of its reaction with the basic substances in the soil. Acidic, sandy, leached soils such as those of coastal plains often are depleted to deficient levels of copper.

Functions of Copper in Plants

Copper is associated with functioning of enzymes. It has roles in photosynthesis, respiration, and processes involving reactions with molecular oxygen.

Effects of Copper on Plant Growth and Quality

Most soils have adequate copper for nutrition of crops. Deficiencies occur most commonly in organic soils and in leached, sandy soils that have the appearance of acid-washed sand. Imbalances of copper in relation to other metallic micronutrients and phosphorus have been suggested to contribute to copper deficiency in crops. Plants absorb about a pound of copper per acre per year. Copper deficiency appears first on the young leaves, especially on young plants, but the symptoms may persist throughout the growing season. The leaves that are affected have general loss of chlorophyll with some striping occurring near the base of the leaves. Die-back and witches' brooms (multiple buds) occur on fruit trees. Bark on trees may become rough and blotchy accompanied with swelling and oozes of brown, water-soluble gum. Fruits may also show these gummy substances. Vegetables show chlorosis or blue-green colorations of young leaves.

Fertilizers

Generally, unless soils are organic or sandy and highly leached, fertilization with copper is unneeded. To supply copper, organic growers should use plant residues, composts, and manures generously. Conventional growers use copper sulfate or copper chelates applied directly to the soil or as foliar sprays. Some fungicides contain copper and can provide copper to plants directly through the foliage or through the roots after the fungicides wash off the plant and onto the ground. Some certifying organizations allow use of copper sulfate or copper hydroxide as a fungicide. Perhaps, by the same reasoning, copper sulfate or hydroxide could be used by organic growers to supplement additions of plant residues, composts, and manures. As with other micronutrients, if deficiencies are diagnosed, copper-containing fertilizers, other than nitrate or chloride salt, are permitted by certifying organizations.

MANGANESE

Manganese in Soils

Manganese is an abundant element in the crust of the earth. Manganese in the soil ranges from 40 to 12,000 lb per acre (top 6 inches). A common occurrence is

5000 lb Mn per acre. Most of the manganese in soil is precipitated as manganese dioxide (MnO_2), which is not very reactive and is nontoxic. For crop nutrition, plants absorb from the soil solution a few pounds of manganese per acre per year. Deficiencies of manganese for crop production might occur in sandy soils from which the element has been leached, in alkaline soils in which the element is precipitated, and in organic soils in which the element is cheleted or from which the element has been depleted by leaching. The solubility of manganese in soils varies with soil acidity. In acid soils, below pH 5.0, the concentration of manganese in soil solutions may be toxic to sensitive crops. Manganese toxicity along with aluminum is considered a major factor limiting growth of crops in acid soils. Only the manganese that dissolves in acid soils has the potential to become toxic. Liming of soils precipitates the soluble manganese and alleviates the toxicity. Above pH 7.5, manganese might be deficient due to the low solubility of manganese compounds in alkaline soils.

Function of Manganese in Plants

Manganese was shown to be an essential element in 1922 (McHargue). Manganese is required absolutely for the activity of some enzymes and will substitute for magnesium in the activation of many enzymes. Manganese appears to be required for the oxygen-evolving system of photosynthesis. Plants have about 20 to 200 ppm (mg/kg) manganese in leaves.

Effects of Manganese on Plant Growth and Quality

Most of the deficiencies of manganese are reported for crops grown on organic soils in humid regions. Manganese deficiency was known as *reclamation disease* because of its association with soils that were drained and brought into cultivation. In organic soils, manganese may be held tightly to the organic matter, or it may be depleted by leaching. Liming of acid organic soils often increases the severity of manganese deficiency by further reducing the solubility of what little manganese may be in solution in these soils that characteristically are low in available and total manganese. Manganese deficiency may occur in mineral soils if pH is high or if the soils are very sandy and depleted of manganese by leaching.

Deficiency symptoms appear as interveinal or spotty chlorosis (yellowing) or necrosis (death) of upper leaves of field crops and vegetables. Plants may be generally stunted, and yields will be limited. In oats, this disorder was known as *gray-speck disease* by pioneers. Fruit and nut trees show symptoms generally on the young leaves, but the symptoms may appear on the old leaves under severe deficiencies. Fruits usually show no symptoms other than lower yields and lighter colors.

Fertilizers for Manganese

Most growers usually do not have to make a specific effort to apply manganese to crops, because only organic soils and very sandy, leached, or alkaline soils are likely to be manganese deficient. Plant residues, composts, and farm manures are the best sources of manganese for the organic gardener. Conventional growers may apply manganese sulfate directly to the soil or as a foliar spray. Organic growers are allowed to use chemical manganese fertilizers, other than nitrate or chloride salts, if manganese deficiency is diagnosed.

MOLYBDENUM

Molybdenum in Soils

The concentration of molybdenum in the soil is very low, being of the order of 4 to 20 lb in the top 6 inches of an acre of soil. Crops absorb less than an ounce per acre per year and contain typically 0.1 to 1 ppm (mg/kg) molybdenum in their foliage. Increased yields of crops have been reported with applications of molybdenum as low as 0.1 oz per acre. Although very little molybdenum is required by crops, it is not a highly toxic element to plants. The solubility of molybdenum in soils is affected by soil acidity. Solubility decreases with increasing acidity, and molybdenum may become deficient for crop growth at soil pH below 5.3. Liming of soils generally restores the solubility of molybdenum to sufficient levels for crop production.

Function of Molybdenum in Plants

The essentiality of molybdenum was demonstrated in 1939 (Arnon and Stout). In plants, molybdenum is essential for nitrate reduction, the initial step in the conversion of nitrate to ammonia. Molybdenum also is essential for nitrogen fixation by legumes. Nitrogen fixation is the process by which legumes, in symbiosis with bacteria, convert nitrogen gas from the atmosphere into plant-available nitrogen. Molybdenum-deficient plants may appear to be nitrogen deficient. These plants are in fact nitrogen deficient because of the inability of plants to utilize nitrate nitrogen or atmospheric nitrogen under molybdenum-deficient conditions and poor availability of soil-borne nitrogen. Any crop is susceptible to molybdenum deficiency, but historically legumes, brassicas, and citrus have been the most notable cases of deficiency. In addition to the stunted and pale appearances of nitrogen deficiency, leaves of molybdenum-deficient plants are burnt around the margins with the scorching progressing to cover the entire laminas of old leaves in advanced stages of deficiency. Often, only the midrib will remain, creating a condition called *whiptail*. Accumulation of nitrate at the leaf margins is associated with the scorching and development of whiptail.

During the 1920s, many years before molybdenum was shown to be essential, whiptail of cauliflower had been described. This disorder was observed on cauliflower that was growing in acid soils (about pH 5.3) where potatoes had been grown on Long Island, New York. The soils had been kept acidic so that the scab disease of potato could be controlled. Scab disease is a surface infestation of actinomycetes or multicellular bacteria on potatoes and makes the potatoes unmarketable. This disease does not grow in acid soils. Whiptail of cauliflower was corrected by liming the soil. Liming allowed for the release of the native molybdenum in the soil. These acid soils of Long Island contained enough molybdenum for plant growth, but the molybdenum was unavailable. Unlike other minor elements, the availability of which increases in acid soils, the availability (solubility) of molybdenum decreases with increased acidity. Nearly 20 years passed between the date that it was noted that liming corrected whiptail of cauliflower and the time that the connection with molybdenum deficiency was made.

Fertilizers for Molybdenum

Liming is the recommended organic practice to prevent or to correct molybdenum deficiency. Molybdenum fertilizers are available commercially. From these fertilizers, molybdenum is applied from 0.5 to 5 oz per acre by direct application to soil, by foliar sprays to crops, or by dusting or soaking of seeds with molybdenum-containing dusts, slurries, or solutions. Use of molybdenum fertilizers may be cheaper than liming, but these fertilizers should not be applied to soils that are below pH 5.0 because they are likely to be ineffective in acidic regimes.

BORON

Boron in Soils

Boron contents in the top 6 inches of soil range from about 10 to 200 lb per acre. Most of the available boron is associated with soil organic matter. Since boron may be leached from soils, its deficiency is widespread in humid regions. Acidic, leached, sandy soils and organic soils are naturally deficient. Leaching of soils and crop removal of boron are gradually increasing the incidences of boron deficiency. Crops remove 1 or 2 lb of boron per acre per year, depending on the type and productivity of the crop.

Function of Boron in Plants

The essentiality of boron was accepted in 1926 (Sommer and Lipman). Earlier work had shown that fruit and other plant tissues contained boron and that low application of boron salts had growth-promoting effects and that high application had growth-inhibiting effects. Boron is not a mobile element in plants, and deficiency symptoms are related to this fact. Die-back of stems may occur during deficiency. A number of nonparasitic diseases—splitting, cracking, corking, hollowing, dry rotting, water-soaking—result from boron deficiency. Examples of these disorders are cracked stem of celery, cracked and hollow root of carrot, cork of apple, brown curd, hollow stem of cauliflower and broccoli, crown rot of sugar beet, and heart rot and canker of table beet. The stems of boron-deficient plants may be brittle.

The precise function of boron in plants is not known. Boron appears needed for movement of sugars in plants. Lignification seems to require boron. Pollination and development of seeds and fruits depend on boron.

Fertilizers for Boron

Boron is best added to the organic garden with plant organic matter. Fertilizer sources include various boraxes and water-soluble materials, which can be applied to the soil directly or sprayed on plants. Boron is a fairly toxic element, and care must be taken not to overapply boron fertilizers. Toxicity may arise from high concentrations of boron in irrigation water in dryland agriculture. Boron toxicity in rotations of a boron-sensitive crop with a boron-fertilized crop is unlikely in humid regions because of rapid leaching of boron. The amount of boron needed by crops varies with species and cultivars, but a common fertilization is 2 lb B per acre per year. Some crops, such as modern cauliflower cultivars, may require 5 or 6 lb of boron per acre per year, but this amount would be toxic to most crops.

CHLORINE

Chlorine in Soils

Chlorine is ubiquitous in nature. It exists in soils as soluble salts of sodium, potassium, calcium, and magnesium and can range from 0.5 to over 5000 ppm in the soil solution depending on environmental conditions and fertilization practices. Chloride (Cl^-) is a major anion in saline soils. Most soils have adequate chlorine for plant nutrition. Chloride is not a highly toxic ion.

Function of Chlorine in Plants

Chlorine was confirmed as an essential element in 1954 (Broyer, Carlton, Johnson, and Stout). Much research with improved techniques over previous procedures were needed to cleanse the air, water, and media of chlorine to deficient levels. However, many years previous to its acceptance as an essential element, it was demonstrated that plants grew poorly in solution culture without chloride salts. Because of its prevalence in the environment, chlorine deficiency in nature is rarely detected. It has a function in photosynthesis, apparently being involved in the evolution of oxygen.

Fertilizers for Chlorine

Chlorine is abundant in nature and in many fertilizers, and one generally does not need to apply chlorine to crops. Chloride is relatively nontoxic and may accumulate in plants without harm in most cases.

NICKEL

Nickel in Soils

Nickel is abundant in the crust of the earth and ranges from 5 to 1000 ppm, averaging about 50 ppm in soils. The solubility of nickel in soils varies with soil acidity, with solubility increasing below pH 6.5 and falling above this value. Nickel deficiency in nature has not been identified as soils have the capacity to supply enough nickel to support crop production. Toxicity occurs only if the soils have been derived from rock with high nickel contents with the resulting soil having nickel concentrations exceeding 1000 ppm.

Function of Nickel in Plants

Nickel has a role in the activity of the urease enzyme in plants and is the seventeenth and last-recognized plant nutrient (Brown in 1987). Nickel may have a role in nitrogen fixation plants. Demonstrations of its suggested essentiality required that plants be grown for several generations without nickel so that the nickel content of seeds did not contribute to the nutrition of the plants. Plants require less than 0.05 mg Ni per kg dry weight (0.05 ppm), and no recorded level has been below 0.2 mg Ni /kg for field-grown plants. Fertilization with nickel-containing fertilizers is not practiced.

BENEFICIAL ELEMENTS

Some elements, such as sodium, cobalt, and silicon, improve the growth of plants or are required by some plants but not by all plants. Enhancement of yields or the requirement of an element by a few plants is not evidence of essentiality. The element must be required by all plants and have an identified function in plants to be considered a nutrient. Sodium is reported to increase yields of table beets, but it is not required for their growth. Cobalt is required for nitrogen fixation by legumes, but since legumes will grow on nitrate or other combined forms of nitrogen, nitrogen fixation is not essential for their growth and development, and hence cobalt is not an essential element. Silicon applications have been shown to increase yields of sugar cane, sugar beet, and rice. Further research may show that some of the elements that are now classified as beneficial will become known as plant nutrients. Until about 1987, nickel was classed as a beneficial element, but research demonstrated that it was required for all plants. Hence, nickel is now classed as a plant nutrient. Among other elements that are suggested as being beneficial are aluminum, vanadium, selenium, and vanadium.

4 Management of Farm Manures

The annual production of manures from livestock on farms in the United States is about 1.2 billion wet tons (about 175 million dry tons). The production of farm manures far exceeds that of sewage sludge (estimated to be about 5.4 million dry tons annually) and is enough to provide nutrients to produce crops on 60 million acres of farm land if it were fully utilized. However, only a small fraction of the potential value of farm manures actually is realized, for much of the manure is wastefully and inefficiently handled. Handling of the large quantities of manure presents major problems to dairy, livestock, and poultry farmers in the nation. These farmers often do not have adequate areas of land on which manures can be applied. In some cases, land is overloaded with manures, but more often, wasteful methods of handling lead to large losses of nutrients and organic matter from manures.

If all of the farm manures produced in the nation could be applied to crop land, many benefits would be realized. The benefits that farmers and gardeners receive from application of farm manures to land are gained (1) through the nutrients that are carried in the manure, (2) through the effects that manures have on increasing the organic matter contents of soils, (3) through the use of manures as mulches, and (4) in the case of farmers with farm animals, from the benefits of having land as a site on which manures can be disposed. Because of the liberal applications of manures to their crop land, dairy, livestock, and poultry farmers will usually maintain the fertility of their land with much more ease than farmers growing only cash crops. It is estimated that the total production of farm manures contains about 7.7 million tons of nitrogen, about 1.9 million tons of phosphorus (P_2O_5), and about 4.2 million tons of potassium (K_2O). If all of these nutrients could be applied to cropland, perhaps about 25% of the nutrient requirements could be met by farm manures. It is estimated also that the organic matter content of farm manures exceeds by at least a factor of two the total amount of soil organic matter that is lost by production of food, feed, and fiber crops in the nation.

Only a fraction of the crop-producing and soil-improving value of farm manures is recovered on crop land. About 60% of farm manures are excreted directly on pastures. The full value of manure excreted onto land is never realized because of poor distribution of the manures on the land and because of great losses of nitrogen from the manures that are left on the soil surface. The fertility of pasture land will decline if the only manures or fertilizers added are those excreted from the animals grazing on the land and if feed is not brought in from off-farm sources.

About 10% of the manure is lost physically through handling, by dropping on roads and lanes, or in piles that are never applied to land. Tremendous losses of nutrients occur through failures to recover the liquid portion, through losses of ammonia by volatilization, and through losses of ammonium, nitrate, potassium, and organic matter by leaching in piles of manures. Phosphorus losses from manures are much lower than the losses of nitrogen and potassium because of the relatively insoluble nature of phosphorus compounds and relatively low amounts of phosphorus in urine. Contrary to the views held by many people, rotting of manure in piles often leads to unnecessary depletion of nutrients and carbon. Generally, composting should be reserved for manures that are very high in bedding. Composting is recommended by organic certifying organizations to allow for pathogens to die and to avoid spreading of pathogens in fresh manure to land.

The manure that reaches land on which harvested crops are grown is only 25% to 30% of the total manure produced, and much of the manure that reaches the land is severely depleted of nutrients. The purpose of this chapter is to present information regarding the handling of manures for conservation of nutrients, for environmentally safe disposal of manure, and for successful production of crops.

COMPOSITION OF FARM MANURES

The value of farm manures with respect to contents of nutrients varies with source and handling of the manures. A ton of average fresh manure, without bedding (hay, straw, sawdust, woodchips, shavings, paper), from large farm animals (cattle, hogs, horses) will have about the same amounts of nitrogen, phosphorus, and potassium as 100 lb of 10-5-10 commercial fertilizer (Table 4.1). Bedding in the manure will dilute its nutrient content. Some manures may have so much bedding, such as wood chips, shavings, or straw, that the manures should be composted before they are applied to the land or additional nitrogen must be added to overcome the potential for nitrogen immobilization if the manures and bedding are incorporated directly in the soil. Fresh manures from sheep may have twice the nitrogen and potassium as manures from other large animals. Fresh poultry manures may have three times as much nitrogen and two or three times as much phosphorus as manures from hogs, cattle, or horses. Manures from large farm animals contain 3 to 5 lb of calcium and 1.5 to 2 lb of magnesium per ton wet weight of manure. Poultry may have much higher calcium

TABLE 4.1
Average Composition of Primary Macronutrients and Water in Farm Manures (Wet Weight Basis of Stored Manures)

	Nutrient Composition, lb per Ton			
Source	Nitrogen (N)	Phosphorus (P_2O_5)	Potassium (K_2O)	Water (%)
Horses	11	5	13	78
Cattle	10	5	8	83
Hogs	12	8	7	87
Sheep	20	10	20	81
Poultry	>30	16	10	35

and magnesium depending on whether the diet of the poultry included limestone or oyster shells, as is the case with laying hens. These manures may be alkaline due to the high levels of calcium carbonate excreted in the poultry feces. Care must be taken to limit the application of these manures to prevent overliming of farmland.

Manures often are marketed as dehydrated or composted products at farm and garden centers. These materials are marketed usually in 50-lb bags. The dehydrated manures are ones from which water has been removed without any rotting of the manures. Marketed dehydrated manures from large animals will have a dry composition of 2-1-2 (%N, $\%P_2O_5$, $\%K_2O$). Dried chicken manures are about 3-1-2. Composted manures have been rotted before bagging and will have a composition of 1-1-1 or 0.5–0.5–0.5. Commercial composted manures also contain considerable water, as much as 25% by weight. In general, the dehydrated manures are the superior product. They have a higher nutrient content and a faster mineralization rate than composted manures. Twenty percent of the nitrogen of fresh poultry manure may be ammoniacal nitrogen.

The better that livestock are fed, the better the quality of the manure. As a rule, about 75% of the nitrogen, 75% of the phosphorus, 85% of the potassium, and 50% of the organic matter of feeds are excreted in the manure. Roughage, if passed through the animals, will be improved in quality as a fertilizer, for the animals act as a composter in removing more carbon than nitrogen. Digestion by livestock removes more of the carbon in feeds than it removes nitrogen, phosphorus, and potassium.

Farm manures contain some of all of the elements that are essential for plant growth, but the nitrogen content of farm manure governs the amount of manure that is applied to land. In general, if enough nitrogen is applied to meet the needs of a crop, the needs for most of the other nutrients will be met (Table 4.2). The phosphorus concentration in farm manures is low, and they are not considered to be adequate materials to provide phosphorus in sufficient total or available amounts in one application

TABLE 4.2
Portions of Total Nutrient Requirements That Will Be Met If Enough Farm Manure Is Applied to Satisfy the Nitrogen Requirement of a Crop

Nutrient	Portion Supplied (% of Total Need)
Nitrogen	100
Phosphorus	25
Potassium	100
Calcium	100
Magnesium	100
Sulfur	100
Minor elements	25

to a crop, particularly on phosphorus-deficient land. The same reasoning applies for minor elements. Generally, another fertilizer should be used to supplement the phosphorus from manures. Often, the supplemental fertilizer is incorporated directly in the manure before application to the land. The practice of mixing phosphate fertilizers with manure gives benefits above those of applying the materials separately. Rock phosphate mixed in at 50 lb per ton of manure enriches the manure with phosphorus, and the manure helps to dissolve the rock and increases the effectiveness of the rock. Minor element deficiencies are rare enough that amendment of manures with minor elements is not a practice. Repeated applications of farm manures to land will enrich the soil with phosphorus and minor elements because these elements will be retained in the soil organic matter added by the manures and in the soil mineral matter, and phosphorus and minor elements are not prone to leaching. Long-term or rates of manure application above the agronomic needs of crops may lead to excessive accumulation of phosphorus in soils. Leaching of phosphorus is a concern in some sandy soils, but leaching is negligible in fine-textured soils. In soils in which phosphorus might leach, it is recommended often that the amount of phosphorus added be the factor limiting application of manures to land rather than using the amount of nitrogen application as the governing principle.

PRODUCTION OF FARM MANURES

The amounts of manures produced by farm animals depend on the type of livestock and on their diets. Approximate productions of manures by livestock are shown in Table 4.3. One horse, one cow, five hogs, or 200 chickens will be about 1000 lb of live body weight.

Farm manures should be incorporated into the soil soon after they are applied to land. If manures lie on the surface for even a few hours after application, nitrogen is lost by ammonia volatilization (Figure 4.1), and the value of manures to increase crop yields is reduced greatly (Table 4.4). Generally, after periods of 2 weeks or more on top of the ground, the value of manures is cut in half. If the manures are never incorporated, their value is practically nil unless the application is very large, that is, in amounts of 1-inch or greater layers or mulches over the soil surface.

TABLE 4.3
Production of Manures by Livestock

	Daily (lb per 1000 lb body wt)			
Source	**Feces**	**Urine**	**Total**	**Total Annual with Bedding (Tons)**
Horses	40	10	50	10 to 12
Cattle	58	22	80	16 to 18
Hogs	50	30	80	16 to 18
Poultry	42	0	42	8 to 10

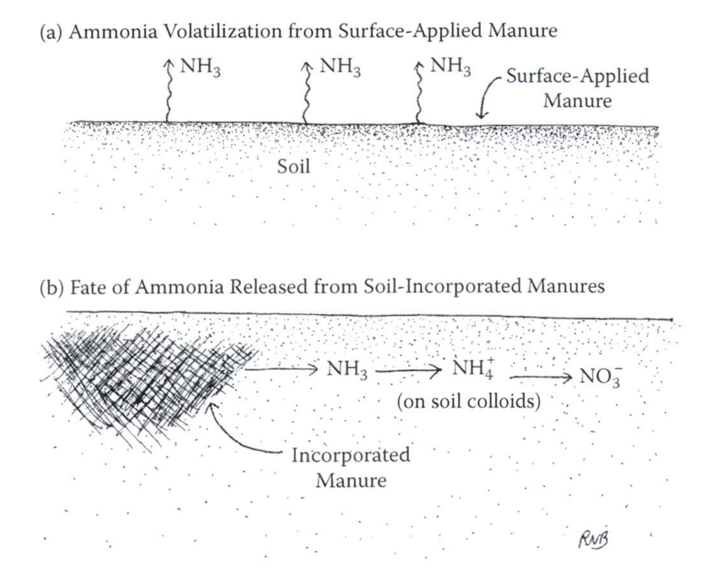

FIGURE 4.1 (a) Loss of ammonia (NH_3) from surface-applied farm manures and (b) retention of ammonia from soil-incorporated manures.

STORAGE OF MANURES

For the livestock farmer and the crop producer, the best handling for conservation of nutrients in manures is to apply freshly excreted manures daily to fields and to incorporate the manures into the soil immediately after application. These practices are not always feasible. Frozen or snow-covered soil may interrupt or prohibit applications. Labor may not be available for daily applications. Manures that must be stored should be piled or left in the barn. Outdoors, manures should be stored on a hard surface if one is available. They should be protected from runoff and away from the eaves of the barn. Potentially 6% of the organic matter, 50% of the nitrogen,

TABLE 4.4
Effect of Time That Manures Lie on Surface until They Are Incorporated on the Relative Value of Manures for Increasing Yields of Crops

Time between Application and Incorporation (Hours)	Relative Value of Manure (%)
0	100
6	85
24	75
48	70
96	55
336 (2 wk)	<50

50% of the phosphorus, and 95% of the potassium could be lost by intensive leaching from piles. These losses can be limited if movement of water through the pile is restricted. A covering of plastic, soil, straw, or hay will help to protect the manure from leaching.

Some dairy and hog farmers liquify manure with a lot of water and store it in lined pits (lagoons) or tanks. These manures are very volatile material once the pits or tanks are disturbed. These manures must be incorporated into the soil immediately after application to avoid huge losses of ammonia and to avoid air pollution from foul-smelling volatile materials.

Farmers should use bedding to soak up and conserve the nutrients in urine. With large farm animals, urine is 20% to 40% of the waste (Table 4.5), but the urine has a disproportionate quantity of nitrogen and potassium. Urine contains 50% of the nitrogen and 60% of the potassium that are excreted by large farm animals. Only about 5% of the phosphorus is in the urine. On an average, failure to conserve the urine results in a loss of about 40% of the nutrients.

Poultry excrete uric acid rather than urea, and all of this material is eliminated with the feces. This process contributes to the high nitrogen concentration in chicken manure, because all of the nitrogen is eliminated in the feces.

Manure that is stored in piles may lose much of its nitrogen and potassium. Nitrogen is lost by leaching with rain water, by gaseous losses of ammonia by volatilization, and by gaseous losses of nitrates by denitrification. Losses of nitrogen are rapid, and losses of nitrogen by volatilization will increase if the pile is turned. For leaching to occur, water must pass through the manure. Leaching losses will vary with whether a cover is on the pile or not, with whether the manure is on a hard surface or frozen at the base, and with the configuration of the pile. A cover on the piles, frozen zones at the bottom of the pile, impervious surfaces under the pile, and high, deep piles restrict movement of water through piles and limit leaching. Losses of nutrients from shallow, flat piles will be very high by leaching and ammonia volatilization.

Volatilization losses of ammonia from the pile may be substantial. Considerable amounts of ammonia are produced in the manure from rapid decomposition of urea and other nitrogenous constituents in the urine and from slow decomposition of the

TABLE 4.5
Distribution of Waste between Feces and Urine and Total Daily Production of Wastes by Farm Animals

	Distribution of Total Waste (%)		
Animals	**Feces**	**Urine**	**Total Daily Production per Animal (lb)**
Horses	80	20	40
Cattle	70	30	70
Hogs	60	40	10
Sheep	65	35	4
Poultry	100	0	0.1

organic nitrogenous compounds of the feces and bedding. Ammonium losses will be restricted from deep piles that are kept moist and that are not too alkaline (below pH 7.2). If piles become dry or are alkaline, half of the nitrogen in the pile can be lost as ammonia in a week or two. Dry piles have lesser capacity to hold ammonia than moist piles, and air movement through dry piles accelerates movement of ammonia from piles.

Some nitrogen is lost by denitrification of nitrate that is produced in the loose, dry, outer portions of piles. The nitrate seeps into the pile, or in barns is trampled into manure on the floor, where nitrate is converted microbially to gases and lost into the air.

Potassium losses occur by leaching in the piles, and the potential for losses of potassium is very high with intensive leaching. Potassium losses generally are less than those of nitrogen, since potassium is not subject to gaseous losses. However, composting of farm manures to produce what is known as well-rotted manures leads to losses of about half of the nitrogen and about half of the potassium. Phosphorus usually is not lost substantially from farm manures in short-term or long-term storage in piles, because phosphorus is not subject to losses by volatilization and due to the low solubility of phosphorus-containing compounds is less likely to be leached than nitrogen or potassium. Manure from piles should be handled as fresh manure and turned under as soon as it is spread to avoid further losses (Table 4.4). Although much of the readily available nutrients and organic matter can be lost from piles, piling is preferable to application of manures to fields without plowing the manures under. In the latter case, losses of nitrogen are very high from volatilization, and half of the nitrogen can be lost in a few days. Proper storage of manure in piles is a recommended practice for conservation of nutrients and organic matter.

TIME FOR APPLICATION OF MANURES

The best time to apply manures is about 1 or 2 weeks ahead of the time that the crop is to be planted. This lead time allows for some decomposition of the manure and lessens the possibilities of plant injuries that often occur with freshly applied manures. Damage that occurs to crops after application of manures is associated with stimulation of plant diseases, with toxicity of ammonium nitrogen present in the manures at the time of application, or with the toxicity of organic substances released from the rotting matter. If feasible, growers should wait 1 or 2 weeks between application of manures to land and planting or seeding of their crops to allow for transformation of the ammonium and organic substances. Also, to avoid toxicities and to lessen the chances of disease infection, growers should not apply fresh manures close to growing crops. Well-rotted manures (compost) can be applied generally at any time and even close to growing plants.

The lead time with fresh manures allows for any ammonium that is present in the manures or that is released from the manures after application to be oxidized to nitrates. Ammonium is the first available form of nitrogen released from the mineralization of organic matter. In the early stages of decomposition of manures in soil, ammonium is released faster than it is oxidized. The chance of substantial ammonium accumulation is high soon after application of manures, particularly if soils are cold, as they might be at planting time in the spring. The toxicity, often referred to as burning, that is encountered with plants that are fertilized with poultry manure is ammonium toxicity. Poultry manures are high in ammonium—as much as 20% of the total nitrogen in fresh manure—and if sufficient time is not allowed between application of manures and planting, substantial damage from ammonium toxicity will occur to seeds or seedlings. Also, when manures are rotting rapidly in the soil, the microorganisms that are carrying out the rotting will attack seeds or roots of seedlings. This problem will occur also with application of highly nitrogenous organic fertilizers, such as dried blood, seed meals, and alfalfa meal. In about 2 weeks time, the easily decomposable materials in the manure or in these nitrogen-rich organic fertilizers will be transformed into nontoxic materials, and the likelihood of damage to seeds or young plants is alleviated. Immature composts, ones that are not stabilized by rotting, also should be applied and incorporated into land about 1 or 2 weeks ahead of planting of crops to avoid potentials of ammonium toxicity or other damages by the raw material.

Decomposition in the soil or compost pile narrows the C:N ratio of manures and bedding and will lessen the likelihood of creating nitrogen deficiency through immobilization that follows application of manures that are strawy or have a high content of other bedding. A 2-week period of decomposition in the soil is not sufficient for manures that have more bedding than feces. These manures should be composted before adding them to the land to avoid immobilization of soil nitrogen. If composting is impractical, additional nitrogen from fertilizers must be applied to the soils with the strawy manures. An application of 1 lb of nitrogen for every 100 lb of strawy manure will counter the effects of immobilization. Again, a lead time of 1 or 2 weeks is recommended between application of manure and planting.

Spring applications of manures are more beneficial for nutrition of crops than fall applications. Fall applications have about half the value of applications made 2 weeks before planting of crops. Applications made in early or late fall have limited benefit because of the leaching of nitrates during the fall, winter, and early spring. Not only do these losses of nitrate lead to limited yields, but the nitrate may contribute to contamination of groundwater. Applications made in late winter or in early spring, again due to losses of nitrate, are not as beneficial as those made near to the time of planting. Generally as previously discussed, applications made 2 weeks ahead of planting give the best results (Table 4.6). These benefits are realized from the conservation of nutrients, especially nitrogen, by the application relatively close to planting and from the degradation during the 2-week lead time of any toxic factors, such as ammonia, that may be present in the manures.

Losses of nitrogen from fall-applied nitrogen fertilizers of any type are usually substantial (Figure 4.2). Growers must be able to purchase fertilizers at half the price of their spring costs to ensure financial benefits from fall applications, and even then, the growers must consider the environmental consequences of losses of nitrogen from fall applications of fertilizers.

The usual function of applying manures to land is to supply plant nutrients. Manures are applied also to increase organic matter in soil or as mulches. With livestock farmers, the principal purpose of application of farm manures to land may be strictly for disposal of the manure. The amount of manure that should be applied

TABLE 4.6
Relative Benefits of Applying Manures in Spring, Winter, or Fall

Time of Application*	Relative Benefit (%)
Two weeks before spring planting	100
Winter or early spring (Dec to Mar)	75 to 80
Late fall or early winter (Nov)	65
Early fall (Sept and Oct)	30 to 50

* Manures worked into the ground immediately after their application.

depends on the purpose of manuring, the quality of the manure, and the crop that is to be supported; however, some general guidelines have been developed for application of manures according to purpose (Table 4.7).

Typically, less than half of the nitrogen and phosphorus and most of the potassium in manures will be released on one season. As little as 10% to 25% of the nitrogen and phosphorus may be available in strawy or poorly handled manures. If the manure is high in bedding—having the appearance of being mostly straw, sawdust, or woodchips—soil nitrogen and phosphorus may be immobilized, thereby creating deficiencies of these nutrients. Manures that have been leached by rain may be depleted of much of their available nitrogen and potassium and will have little value as fertilizers.

An application of 20 tons of good quality manure per acre supplies about 200 lb of nitrogen, about 100 lb of phosphate (P_2O_5), and 200 lb of potash (K_2O). From this

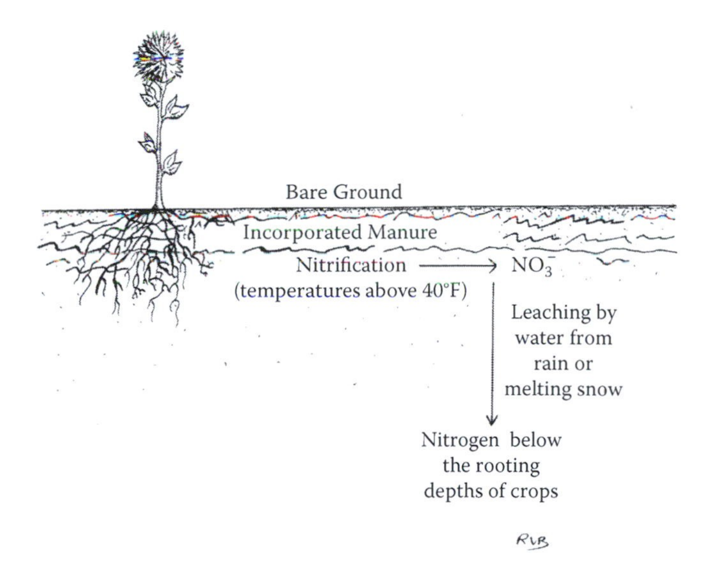

FIGURE 4.2 Loss of nitrate formed from manures applied in fall, winter, or early spring.

TABLE 4.7
Recommended Amounts of Fresh Manures According to Purpose of Manuring

Purpose	Amount to Apply	
	lb/100 sq ft	Tons/Acre
Fertilization	100	20
Increasing organic matter	150	30
Mulching (1.5 inches)	500	100

application, about 100 lb nitrogen, 50 lb phosphate, and 200 lb potash are potentially available. In the season of application, this amount of manuring supplies all of the nitrogen and potassium and part of the phosphorus required by most crops (Table 4.2). High-yielding crops, such as corn for grain, might require higher applications if the soil is of low fertility and if farm manures have not been applied previously or recently. In most cases, however, as far as plant nutrition is concerned, very little economic benefit is achieved from application of manures in excess of 20 tons per acre. The application of 20 tons per acre often is referred to as the *agronomic rate of application.*

Residual effects of manures are considerably lower than the benefits received during the season of application. In the year following application (second year), about 10% to 15% of the nitrogen remaining from the manure is released to crops, compared to 50% released in the first year. In the third crop year, roughly 5% is available, and in the following years 1% or 2% of the remaining nitrogen is available (see Tables 3.1 and 3.2). A continued program of yearly applications of farm manures is needed to maintain soil fertility with crop production.

Increasing organic matter in the soil in a short time requires higher applications of nitrogen than those needed for nutrition of plants. Applications of 30 tons per acre per year often are recommended to increase the organic matter content of soils. Growers should be cautioned that the nitrogen supplied with high applications, those above the agronomic rate, of farm manures may exceed considerably the amounts of nitrogen required by crops and may have unfavorable effects on plant growth and quality and on the environment. Applications in the range of 40 to 90 tons per acre per year should be avoided because of the chances of environmental pollution. Applications of manures at or below the rates required for crop nutrition in the long term will have favorable effects on increasing organic matter in the soil. Since little economic benefit is received from accelerated building of soil organic matter, in most practices, growers should limit applications to about 20 tons per acre per year.

If a grower does not have sufficient quantities of farm manure for recommended applications of 20 tons per acre per year, much smaller applications are beneficial to the land with respect to supplying plant nutrients for a current crop and with respect to increasing organic matter and reserve nutrients in the soil. Nutrient supply would be directly proportional to the manure applied. A 5-ton application will supply 25% of the nutrients of a 20-ton application. A 5-ton application over a long period—30 or 40 years of more—will maintain soil organic matter in land with crop rotations and will greatly slow the losses of organic matter in land cropped continuously to cash grains, corn, oats, and wheat. Because of the low amount of crop residues produced by vegetable crops, small applications of farm manure may not be sufficient to maintain soil organic matter.

Increases in soil nitrogen will occur in proportion to increases in organic matter with manuring, with about a l-lb increase in nitrogen for every 20-lb increase in soil organic matter. Manuring of crop land with less than 20 tons per acre annually usually does not maintain levels of potassium, calcium, magnesium, and phosphorus in the soil. Cropping of land manured with less than 20 tons per acre will lead to depletion of soil of these mineral nutrients unless they are added from some outside source. Manured land will be slightly more acidic than unmanured land. Liming to raise pH will maintain the calcium and magnesium.

A mulch should be about 1.5 inches thick with well-rotted manure. Crops generally will not be overnourished from surface-applied manures applied as mulches of this thickness. Plants receive the nutrients that leach from the mulches. Half of the nitrogen might be lost to the air, but enough remains to nourish the crops. Well-rotted manures (composts) are selected in preference to fresh manures for mulches. Well-rotted manures are desirable because they are nicer to handle than fresh manures and give no offensive odors. Ammonia volatilization from well-rotted manures will be much less than that lost from surface-applied fresh manures. Well-rotted materials can be placed close to plants without likelihood of damage to plants. Strawy manures also make good mulches. Immobilization of nitrogen is nil for these mulches if they are left on the surface of the ground and not soil-incorporated during the growth of a crop.

OBJECTIONS TO MANURES

Fresh manure, especially cattle, horse, hog, and chicken manure in increasing order, is a favorable breeding place for house flies. Offensive odors may be emitted from manures. Flies and odors may be controlled somewhat by piling manures in compacted heaps. Methods of storage that emphasize minimizing losses, usually help to attend to problems with odors. Piles and lagoons that are opened for spreading of manures to fields will stink. Manures should be incorporated into land as soon after application as is possible to keep down offensive odors as well as to restrict losses of nitrogen.

Weed seeds in manures are a serious problem. Heavily manured land often becomes heavily infested with weeds. The weed seeds may come from hay that is fed to livestock or from the bedding that is mixed with the manure. Mature weed seeds may be in the hay that is wasted and packed into the manure, and the mature seeds of the hay feed as such may produce weeds in another crop in which the volunteer forage plants are unwanted. Weed seeds may pass though large farm livestock without much reduction in seed germination. Seeds may be destroyed in the gizzard of poultry but not reliably. Fermenting and heating (composting) of farm manures is an inadequate practice to reduce weed populations even though composting often is recommended as a method to destroy weed seeds in farm manure and in plant residues.

Farm manures might contain pesticide residues, depending on the feeds for the livestock and poultry. Hogs are often fed diets of copper or zinc salts to increase the rate of gain of the hogs and to kill intestinal worms. Composting will destroy organic materials, such as antibiotics fed to livestock, but the inorganic metallic salts will be unaffected or possibly increased in concentration by composting of manures. Organic antibiotics also will be degraded by microorganisms in the soil and are not considered as a problem limiting the application of manures to cropland.

Manures can be high in ammoniacal nitrogen. The delay of a few days or 1 or 2 weeks between application of manures to land and seeding or planting of crops helps for ammonium nitrogen to be oxidized to nitrate thereby reducing phytotoxicity from the ammonium. The delay in planting also allows for microbiological activity to become stabilized after application and helps to lessen chances of rot diseases affecting crop seeds or transplanted seeding. Some certifying agencies alert growers to the possibility of high levels of nitrates accumulating in vegetables that are fertilized with raw manures. It is difficult to give adequate nitrogen nutrition from fertilizers or from soils to some vegetables without having nitrate accumulation. Ammonium toxicity from raw manures should be much more of a concern than nitrate accumulation in vegetables.

Some human pathogens, such as salmonella bacteria, might be transmitted by application of raw farm manures to land. Composting of the manure or application of the manure to the land well in advance of anticipated harvests of food crops is recommended to lessen chances of transmission of pathogens from manures to humans or livestock.

Composting of farm manures is recommended by organic certifying organizations. Composting can reduce the concentrations of pesticides that are organic chemicals but will not eliminate the zinc or copper that might be added in hog diets. Composting will kill pathogens of plants, livestock, poultry, and humans. In lieu of composting, some certifying organizations recommend a period of 4 or more months between application of manures and harvest of crops for human consumption, thereby allowing for sheet composting in the soil.

5 Composting

Composting is the controlled rotting of organic matter. The purpose of composting is to convert organic material that is unsuitable for incorporating into the soil into a material that is suitable for mixing with the soil. The main drawback of uncomposted organic matter for land application is its wide carbon to nitrogen (C:N) ratio. Incorporation of organic matter with a wide C:N ratio is likely to stimulate immobilization of available nitrogen and other plant nutrients in the soil. Immobilization is the consumption of available soil nutrients by microorganism that are rotting organic matter in the soil. Immobilization occurs when the organic matter being decomposed does not contain enough nutrients to feed the microorganisms that are doing the rotting; thus, the organisms obtain the nutrients from the soil. Immobilized nutrients are unavailable to plants until they are released from the cells of the microorganisms that consumed them.

Among materials that have wide C:N ratios and that should be composted are sawdust, woodchips, bark, straw, paper, dead leaves and needles of trees, dead garden residues, cotton cloths, and manures with a lot of bedding. During composting, carbon is lost more rapidly than nitrogen is lost from the organic materials, with the result that the C:N ratio of the final product is much narrower than that of the original organic materials.

Another potential but limited benefit of composting, other than narrowing of the C:N ratio, is to destroy weed seeds and pathogenic plant diseases. Weed seeds may rot or become unviable during composting. Diseases may be killed by heat generated by composting, by competition from other organisms, or by microbial toxins generated during composting. However, destruction of weed seeds and diseases during composting cannot be assumed. Weed seeds on the outside of the pile will not be rotted. Weed seeds may be carried in by wind or birds or by adding of mature plants to the pile after it is almost finished composting. Weeds at the perimeter of the pile may produce seeds that infest the pile. Composts that are not matured fully, that is, rotted fully, may contain weed seeds from plant materials that were the original material added to the compost pile. Composts made from yard wastes, hays, and farm manures often contain viable seeds of weeds from the original materials. Mature composts should be covered to protect them from infestation with weed seeds from the surrounding areas. Weeds may be controlled by covering the pile with black plastic to prevent the growth and maturation of any plants that start in the pile from the germination of weed seeds.

The high temperatures (130°F or higher) in the internal region of the pile may kill some plant diseases caused by fungi or bacteria. Plant disease organisms and nonpathogenic organisms compete for the same substrates during composting; the nonpathogenic organisms may out-compete the diseases leading to death of the diseases by starvation. Organic acids and other compounds generated during composting can be toxic to diseases. However, as a safety precaution, diseased material should not be added to the pile at all. Diseases that are not exposed to the inside heat of the pile will not be affected by composting. The pile may not heat up enough after the first or subsequent turnings to kill diseases of plants that were originally on the outside of the pile. Competition and toxins in the pile may not be uniform to kill all diseases below levels that might infect crops. Viral diseases are seldom killed during composting due to insufficiently high temperatures in the pile.

During composting, half or more of the bulk of the pile is lost by decomposition, waste, or unavoidable losses. At least half of the carbon is lost during composting. Loss of carbon is a mandatory process for successful composting, because one of the goals of composting is to narrow the C:N ratio. Microorganisms use the organic material as a source of energy. Respiratory production of carbon dioxide, which escapes from the pile, represents most of the loss of carbon. Nitrogen is lost by volatilization of ammonia and leaching and denitrification of nitrate. However, carbon, as carbon dioxide, is lost faster than nitrogen, and as a result the C:N ratio is narrowed. Potassium is lost by leaching. Phosphorus is conserved, because it is mostly in sparingly soluble compounds that are neither leached nor lost as gaseous substances. Finished compost is a dilute fertilizer having a composition of about 1-1-1 (%N - $\%P_20_5$ - $\%K_2O$), but varying slightly with the original materials that were incorporated into the pile and with regard to handling of the compost during and after the process of composting.

As a rule, composts with less than 1% total N should not be incorporated into the soil without supplemental fertilization with a readily available source of nitrogen. These composts have too low a concentration of nitrogen to benefit crops and may lead also to N immobilization. Composts with less than 1% N can be used as mulches without concern about nitrogen immobilization in the soil. The C:N ratio of finished compost may vary from about 10:1 to 50:1, varying with the nature of the material being composted and with the maturity of the finished compost. Fully mature composts generally have narrower C:N, about 15 to 20:1, than raw, unstable, or unfinished composts. Finished, mature compost is slightly alkaline, being about pH 7.5. Acidity or alkalinity of a compost is a good determination of maturity, with immature composts being acidic (pH 5 to 6) and mature composts being slightly alkaline.

Decisions have to be made as to whether to compost materials or to add them directly to soil. If the bulk of the uncomposted organic materials has a C:N of about 50:1, it need not be composted in a pile and can be added directly to the soil. The C:N ratio will soon narrow by sheet composting in the soil so that nitrogen will be released. Nitrogen fertilizers can be added to the soil with the organic matter to narrow the C:N ratio further and to stimulate rotting. To avoid toxicity from the nitrogen fertilizers, growers should be cautious about trying to narrow the C:N ratio of ratios wider than 50 with one application of nitrogen or without waiting for a week or two for the material to stabilize after application in the soil. If the available amounts of materials with narrow or moderately narrow C:N (less than 35:1) ratios are small in volume or mass, these materials can be used in composting to accelerate the rate of composting of materials with wide C:N ratios (greater than 35:1).

BENEFITS OF APPLICATIONS OF COMPOSTS TO LAND

Composts are added to soils to improve soil fertility by modifications in soil chemical, physical, and biological properties. Composts are dilute fertilizers, and chemical improvements include increases in the reserves of plant nutrients, especially nitrogen, in the soil. The capacity of soils to retain nutrients from the composts or other fertilizers is increased after additions of composts because of the increases in the cation exchange capacity of soils, which is the capacity of soils to hold positively charged ions, such as NH_4^+, K^+, Ca^{2+}, and Mg^{2+}. Soil organic matter (including humus, a stable form of organic matter in the soil) has a higher capacity for cation exchange than clays. The water holding of compost-amended soil increases as the organic matter in the soil increases. Nutrient retention will be improved by restrictions in leaching in compost-amended soil.

Physical changes in soil include decreases in the bulk density, increases in the water-holding capacity, and better structure or soil tilth. These changes result from the increases in organic matter imparted by the compost and from the effects of organic matter on soil physical properties. In the short term, the beneficial effects of adding compost are related directly to the nutritional value of compost, whereas in the long term, the benefits result from the increases in humus.

The effects of composts on soil organisms are complex. Composts are a nutritional base for microorganisms and earthworms and other macrofauna and will stimulate their growth in soils. All composts added to the soil should be mature composts; that is, the composting process should be finished in the pile and not in the soil. Immature composts often are high in ammonium or unstable nitrogenous compounds that are mineralized rapidly. High ammonium concentrations in soils are phytotoxic and may affect macrofauna as well as the crops. Raw composts with undecomposed, unstable organic matter may stimulate plant pathogens, such as damping off disease of seeds and seedlings and root rots. However, well-matured composts may have suppressive effects on plant pathogens through microbial competition with the pathogens for nutrients and carbon or through specific parasitic toxicity on or colonization of the pathogens by beneficial microorganisms.

A detrimental biological effect of composts is imparted through their increasing weeds in the soil. Composts may carry weed seeds and infest the soil. Also soil-borne weeds benefit from the enhanced soil fertility in composted land and may outperform the crops in response to the compost, especially in the first year of compost application.

Composts frequently are used as a substitute for peat moss in potting media. Only well-matured composts should be used in potting media. Raw materials may upset the nutritional relations of plants growing in the media, may be phytotoxic, and may stimulate diseases. Even mature composts are more unstable than peat moss in a medium. Composts will rot more rapidly than peat and may have to be used in higher proportions in the medium than peat. Also, compost-based media will decrease in volume and mass as the compost decomposes.

PROCEDURES FOR COMPOSTING

Procedures for composting involve collecting the materials, selecting the site, and selecting the process for composting. For production of composts in small-scale operations for use on farms and gardens, the grower does not have to be precise in following these procedures unless the production of the compost is urgent. However, in large-scale and commercial operations strict adherence to procedures is necessary to obtain a good product in a timely manner.

Selecting the Materials

A substantial amount of material must be collected to build a compost pile. The ideal size of a pile for home composting is 5 ft high x 5 ft wide x 5 ft long. If enough material cannot be collected to build a pile of these dimensions, efforts should be made to make the pile as high as possible. Flat, pancake-like piles will not compost well and will lose more nutrients than deeper piles. The length and width of the pile can exceed these dimensions considerably, but the height of the pile should not exceed 8 to 10 ft. Piles above this height may become too hot in the interior and kill the composting organisms. Ignition by spontaneous combustion has occurred in deep piles. Piles somewhat deeper than 5 ft will be more active in cold weather than shallower piles due to the insulating effect of the deeper piles. Use of sewage is not permitted in organic farming.

Materials for composting are divided into two classes: *carbonaceous* materials and *nitrogenous* materials (Table 5.1). Carbonaceous materials have wide C:N ratios (e.g., C:N > 100:1), and nitrogenous materials have narrow C:N ratios (e.g., C:N < 35:1). A mixture of carbonaceous and nitrogenous materials is used frequently in the construction of a compost pile. Woody and nonwoody materials may be used together but should be kept separate if possible. Woody materials will take much longer to rot than the nonwoody materials. Inclusion of woody materials in the compost pile may prolong unnecessarily the duration of composting. Paper can be used but should not exceed 10% of the volume of the pile. Waxed paper and colored paper should be avoided. Waxed paper will be resistant to decay, and colored paper may contain heavy metals.

The carbonaceous materials provide the bulk of the organic matter to be composted, and the nitrogenous materials accelerate the rate of composting. Generally, the

TABLE 5.1
Examples of Common Carbonaceous and Nitrogenous Materials for Compost Piles

Carbonaceous Materials		Nitrogenous Materials	
Sawdust	Garden residues	Green leaves	Wool cloths
Wood chips	Tree leaves and needles	Legume hay	Fur, silk
Bark chips	Straw	Coffee grounds	Soil
Grass hay	Paper	Tea bags	Manures
Cotton cloths	Nut shells	Grass clippings	Sewage
Yard trimmings (non-green)		Spoiled vegetables	

narrower the ratio of parts of carbonaceous materials to nitrogenous materials, the faster the compost will be finished.

Nitrogen fertilizers may be used instead of organic nitrogenous materials. Fertilizers are added usually by sprinkling them over a foot-deep layer of organic matter during construction of the pile. Ammonium-based fertilizers are better than nitrate-based fertilizers. The microorganisms that are carrying out the composting utilize ammoniacal nitrogen in preference to nitrate nitrogen. Suitable fertilizers would be urea, ammonium sulfate, ammonium nitrate, seed meals, dried blood, manures, and sewage from nonindustrial sources. Strict interpretation of organic practices may deter use of urea, ammonium sulfate, and ammonium nitrate. Liberal interpretation of organic practices will allow their use, because these chemical forms of nitrogen will be converted into organic forms by the biological processes of composting.

Soil is added often as a nitrogenous material. Soil also adds some weight to the pile and helps to hold it in place. Soil is also a source of inoculum of microorganisms that perform the composting. The soil should be the richest available. Soil that is dug from underneath an old compost pile or from a barn or barnyard will be rich in nitrogen. The amount of soil added to the pile should not exceed one-half inch to 1 inch for every foot of compost in the pile. Too much soil will make the pile heavy and possibly anaerobic in wet weather. The finished product with an abundance of soil added will have the characteristics of soil rather than of organic matter.

Phosphorus fertilizers can be added to the pile to fortify the compost. Rock phosphate, bonemeal, and superphosphate (not organic) are suitable. The acids formed during composting will aid in solubilization of rock phosphate, converting some of it to a material that is similar to superphosphate, which is manufactured by acid treatment of rock phosphate. Granite dust or greensand added to enrich the pile with potassium will not be effective due to their very sparingly soluble nature even in acidic environments.

Many procedures for composting call for adding of lime or wood ashes to the pile. Adding of these materials should be avoided. The increase in alkalinity from their addition will accelerate the loss of ammonia by volatilization and will lower the value of the final compost. Compost piles undergo changes in pH as composting proceeds. The initial reaction of the pile will be acidic, but this development of acidity is part of a natural process. A particular group of thermophilic microorganisms adapted to acidic conditions utilize the easily decomposable fraction (sugars, starches, organic acids, proteins) of the organic matter and generate acidity in the process. These organisms die out as the composition of the residues changes. Another group of microorganisms then becomes dominant in the process of composting. These organisms use different substrates and have different effects on pH than the original group. Nothing needs to be done to alter this sequence of events. Finished compost is slightly alkaline, perhaps about pH 7.3 to 7.5. Lime or wood ashes should be added to the soil to which compost is applied, not to the compost directly. Some commercial composts made from municipal solid wastes, particularly biosolids (sewage sludge), often are limed heavily to control disease organisms. These composts may have pH values of 8.0 or more. These composts might be considered as limes rather than fertilizers.

Coal ash or coal-based charcoal ash should not be added to the pile because of the potential phytotoxicity of components in coal.

Bones, meat, and grease should not be added to the pile. Bones will not compost unless they are ground. Bones and meat will attract animals to the pile, but if the compost pile is enclosed and inaccessible to vermin, meat might be considered to be a compostable material. Meat would be a nitrogenous material. Grease may be slow to compost and adds little besides carbon.

Selecting the Site

The site should be handy to the land on which the compost is to be used. A nearby source of water is desirable. Rainfall may not provide adequate moisture to initiate and maintain composting. A partially shaded location helps to keep the pile from drying. Windbreaks will help to prevent drying. If the compost pile is in a residential neighborhood, some screening from view is desirable. Compost piles are likely to emanate odors and should be located remotely enough to avoid odors drifting into inhabited zones.

Pits are not necessary for composting unless the climate is very dry or cold. In these cases, composting in pits will help to prevent drying and loss of heat from the pile.

Selecting the Process

The selection of a process is dictated by the time and materials that are available. For producing compost for the current season, growers should consider the *14-day process*. If time to produce the compost is not a factor, the grower will want to use the *90-day* or the *long-term* procedure. If the amount of material to be composted is great and labor is short, the grower may opt for *sheet composting*.

Fourteen-Day Process

This procedure is used when time is short, such as in the spring, and when a source of finished compost is not available. This process is sometimes called the *Berkeley process*, after Berkeley, California, where the process was developed by engineers who wanted a process for rapid composting of municipal solid wastes, yard wastes, and leaves.

High-quality organic materials are needed to ensure rapid composting (Table 5.2). A mixture with a ratio by volume of two parts carbonaceous materials to one part nitrogenous materials is ideal for rapid composting. A mixture of farm manure and bedding (hay, straw) directly from the stables gives about the proper ratio of materials for the 14-day process. For rapid composting, sawdust and wood chips in bedding should be avoided, for they will be too resistant to decay. With most materials, the 14-day process is labor intensive. The materials have to be shredded and moistened as the pile is constructed. Shredding of the materials increases the surface area and accelerates rotting. Shredding is difficult without power-driven equipment. Working with power-driven equipment can be dangerous, and the equipment is expensive. Increasing the fraction of nitrogenous materials and using carbonaceous materials

TABLE 5.2
Selection of Material for Various Composting Processes

Process	Material	Objective
14-day	High-quality compostable solid wastes. Two parts carbonaceous wastes to one part nitrogenous wastes by volume.	To make compost for the current season.
90-day	All kinds of compostable solid wastes. Three parts carbonaceous wastes to one part nitrogenous waste by volume. Wider ratios can be used but will prolong process.	To make compost for the next season or for use late in the current season. This process can produce compost for use at any time in warm climates.
Long-term (1 or 2 years)	All kinds of compostable wastes. No restriction on ratios of carbonaceous wastes to nitrogenous wastes.	To make compost with no time schedule.
Sheet	All kinds of compostable wastes. No restriction on ratios of materials. The more carbonaceous the material, the longer the time needed for composting.	To produce compost on the site on which crops are to be grown and to compost a large amount material with a low input of labor. To allow for composting while the crop is growing.

that are already finely divided or soft (needles, leaves, fine garden residues, trimmings or spoiled vegetables, manures with hay or straw bedding) may enable elimination of the step of shredding.

A pile constructed by the 14-day method should heat up in about 2 days. Warming of the pile indicates that microbial action is occurring. If it does not heat, insufficient water or insufficient nitrogen may be available for rapid composting, and these materials will have to be added to accelerate the rate of composting. The pile must be turned every 3 or 4 days for about 10 days. Turning of the pile aerates it. Usually, no additional turnings are needed after the third one. Failure to have sufficient water or nitrogen or failure to turn the pile on time will delay composting. After about 14 days—sometimes 21 days or longer depending on the success in constructing the pile, the climate, and frequency of turning—the organic material should be decomposed. At this time, some of the original material may be recognizable, but it should fall apart easily on handling. Slick and glossy materials have not composted sufficiently. If some of the compost pile appears to be finished and some not finished, the compost may be screened to separate the finished product from the unfinished material. Chicken wire or about 1-inch-mesh hardware cloth may be used to construct a screen. The material that passes through the screen may be ready to use. It should be examined to determine that it is indeed compost and not just finely divided material that still requires further rotting. Finished compost should be black and crumbly and with an earthy odor. The material that does not pass through the screen should be added to another compost pile.

Ninety-Day Process

This process is sometimes referred to as the *Indore process*, being named after a region of India where considerable research on the process occurred. Lower quality materials can be used in this method than with the 14-day process (Table 5.2). Shredding helps to accelerate composting but is not necessary. A volume ratio of three parts carbonaceous materials to one part nitrogenous materials is a typical mixture of organic matter. Wider ratios may be used, but the time for composting will be lengthened. Times of composting exceeding 120 days are common, even though this process is called the 90-day procedure. Compost made by this process normally is not available for use in the same year in which the construction of the pile was initiated in temperate climates. Storage and protection of the finished compost is necessary if it cannot be used in the current season.

Construction of the pile may begin with a foundation of coarse stems or twigs (Figure 5.1). These materials should not exceed 1 inch in diameter, and the layer should be about 6 inches thick. The organic materials, soil, and optional fertilizers are layered on this foundation. Layering of materials is optional but helps to measure material and to keep the proper proportions of carbonaceous materials, nitrogenous materials, and soil. The subsequent layers after the base are commonly about 12 inches thick. The substrate (sometimes called *feedstock*) can be added as it becomes available. Adding material over a period of time usually lengthens the time needed to get finished compost. If raw materials are added continuously to a pile, the compost may never be matured.

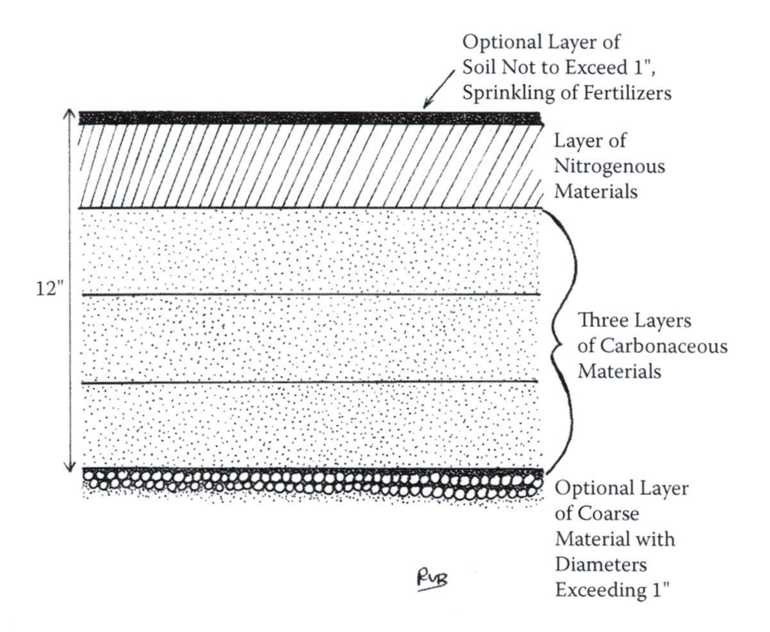

FIGURE 5.1 Construction of compost pile for 90-day process. The purpose of layering is to measure volumes of materials. After first turning, layers are intermixed.

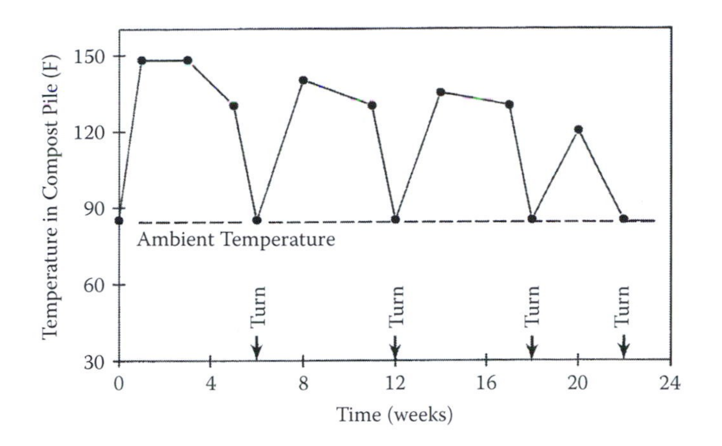

FIGURE 5.2 Cycles of heating and cooling of composts during the 90-day process.

The pile may be watered as it is constructed, but watering is not as essential for the 90-day process as for the 14-day process. Most workers rely on rainfall to provide water for the 90-day process.

In a week or so after construction of the pile, it should heat up to 130 to 150°F. Small piles only about 3 feet high or in small composting containers will not heat to this level because of cooling by radiation from the small mass of material. After 5 or 6 weeks, the pile will cool to ambient temperature (Figure 5.2). At this time, the pile should be turned to aerate it. The turning should be repeated each time that the pile returns to the ambient air temperature. Sometimes the piles are not turned, but not turning the pile usually adds to the length of time required for composting. Instead of turning, holes may be poked in the pile with sticks or bars to give channels through which air may enter easily. Sometimes poles or boards are placed vertically in the pile, also providing channels permitting air to diffuse into the pile. In some large-scale operations, air is pumped or sucked into the piles through perforated pipes buried underneath the piles. This practice provides good aeration and generally shortens the amount of time required for composting relative to turning and saves labor.

Piles that are not turned or provided with forced aeration likely will take more than 1 year for completion of composting. Labor savings are considerable if a pile is not turned. Nitrogen is conserved in the pile if it is not turned. Each time that a pile is turned, some ammonia escapes into the air. Composts from piles that are not turned may have 10% to 20% more nitrogen than turned piles, that is, 1.1% to 1.2% N compared to 1.0% N.

After about the third turning, the temperature of the pile may not rise above the ambient temperature, perhaps indicating that the pile is composted. The compost should be examined to ensure that it is finished. The original organic matter will not be recognizable in the finished product. The organic matter should be black and crumbly and have an earthy odor. If it does not fit these criteria, the compost may not be finished. Shortage of water in the pile is a likely factor that causes failure of the pile to heat up when it is turned.

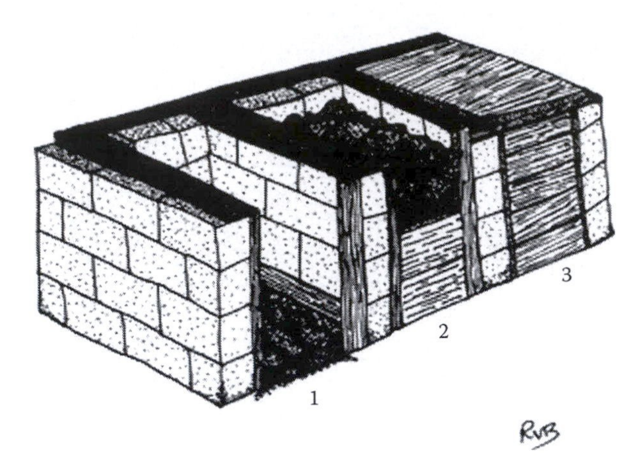

FIGURE 5.3 A three-bin composting unit. Bins 1 and 2 are for receiving and turning of compost. Bin 3 is covered and is for storage of finished compost.

The finished compost can be incorporated directly into the soil. If the season is not proper for use of the compost, the compost should be stored by covering it with soil, hay, straw, or black plastic. These coverings will help protect the finished compost from leaching by rain and may help to keep weeds from growing in the pile. Small amounts of composts may be stored in closed bins. Large bins for storing compost or even for working of composts can be constructed from snow fencing, lumber, or concrete blocks (Figure 5.3).

Sheet Composting

This process involves composting directly in the soil on the site on which crops are to be grown. Sheet composting often is used in large-scale operations involving large acreage or large amounts of materials. It is a valuable process when labor is short. Considerable equipment may be needed in substitution for labor. Turning under green manures or residues from a previous crop on gardens or fields is a form of sheet composting.

Organic materials are worked into the soil 6 or 8 weeks in advance of date of planting of the crop. This lead time is needed in order for the organic matter to decompose sufficiently so that immobilization of nitrogen is not as severe a problem as it would be if the crop were planted soon after the organic matter is incorporated. Nitrogen deficiency in the crop can be a severe problem resulting from sheet composting, and often additional nitrogen fertilizers are added to ensure that nitrogen deficiency does not occur. Try to add about 1 pound of nitrogen for every 100 pounds of organic matter added.

Soils in which sheet composting has taken place may be dry. The coarse organic matter gives a lot of open channels through which water may evaporate from the soil into the atmosphere. A poor seedbed may result from sheet composting if the organic matter has not decomposed sufficiently before planting of crops. The longer the interval between application of organic matter for sheet composting and the planting of a crop, the better the condition of the seedbed.

Commercial Composts

Many kinds of composts are marketed or given away. Composted farm manures commonly are offered for sale. Commercial composts of farm manures are excellent products and can be added to soils liberally. Sometimes farm manures are cocomposted with other agricultural wastes depending on the agricultural activity in the area; for example, composts are made from chicken manure and fruit pomace. These cocomposted materials are excellent for providing nutrients to crops.

In municipal areas, yard wastes (grass clipping, chipped twigs and branches, leaves, or whatever else that may be raked up) are composted and sold or given away. These composts generally are good for incorporation into soils. They often are lower in nitrogen than the agricultural composts. Composts of autumn leaves are a common product in high demand. These composts are quite low in nitrogen, perhaps only 0.7% N in fresh composts, and they are not good as fertilizers. In fact, supplemental fertilization is needed to compensate for the low nutrient value of the composts and to promote plant growth in soils amended with leaf composts. Well-matured leaf composts with 1 year or longer of decomposition are better than those of lesser maturity, but supplemental fertilization with nitrogen is still recommended.

In some areas, composts are made from biosolids (sewage sludge) that are cocomposted with wood chips or other municipal solid wastes (paper, for example). These composts are rich in nitrogen—often as much as 3% N in fresh mature compost less than 1 year old. The high nitrogen concentration makes these materials a valuable fertilizer. Frequently, composts less than 1 year old have considerable ammonium contents. Seeds or transplants into soil freshly treated with this compost may be injured by the ammonium. After application of compost to land, the ammonium, and the toxicity, quickly dissipates by nitrification or by volatilization within a few days to a week. So an interval of one week between application and planting is recommended to avoid inhibition of seed germination or injury to transplants.

Concern exists about the contents of cadmium, zinc, copper, lead, and other metals in composts from municipal solid wastes. Much care is taken currently to avoid use of materials that would add these metals to the compost. Composts offered for sale or given away by municipalities are stringently governed by federal and state regulations. Only composts that are safe to apply to land without governmental restrictions should be used for agricultural purposes. All growers who use composts made from biosolids should ask for a report of the chemical analysis of the composts and compare these results with the federal standards for permitting use of biosolids on land. Organic certifying organizations prohibit use of biosolids-based composts due to the perception that these composts have unwanted metals.

6 Management of Green Manures

FUNCTIONS OF GREEN MANURING

Green manuring is the tilling of green, growing plants into the ground to improve the condition of the soil. Green manures improve soils in several ways:

1. Addition of organic matter
2. Addition of nitrogen with legumes
3. Conservation of nutrients
4. Protection of soil against erosion
5. Biocontrol of diseases

ADDITION OF ORGANIC MATTER

Increasing the organic matter content of soil is the primary purpose of green manuring. Turning a green manure into the soil adds 1 or 2 tons of dry matter per acre and is the surest method of increasing soil organic matter. Although some effects on soil conditioning, such as aggregation, may last for several years, most of the benefits occur in the growing season following incorporation of the green manures. Green manures decompose rapidly due to their succulence, and about 75% of the organic matter will be rotted in the first season, leaving little residual effects afterward. For maximum benefits, the green manure crop should be in a crop rotation every third year, such as a rotation of corn-oats-green manure or vegetable-vegetable-green manure. Green manures do not have much effect on increasing soil organic matter unless the crop is grown for at least 1 year or growing season. Growing the green manure for more than 1 year usually is undesirable, because this practice takes the land out of production for too long. A winter *cover crop* (see below) seeded in the fall and turned under in the early spring may not add much organic matter to the soil. Cover crops that are turned under late in the spring after at least 12 inches of growth have occurred may increase organic matter in the soil. The primary purpose of the cover crop is to protect the soil against water and wind erosion during the seasons when the ground otherwise would be bare.

ADDITION OF NITROGEN

Nitrogen is added to the soil by *nitrogen fixation* if the green manure crop is a *legume*. Nitrogen fixation is a process by which nitrogen in the air is converted into

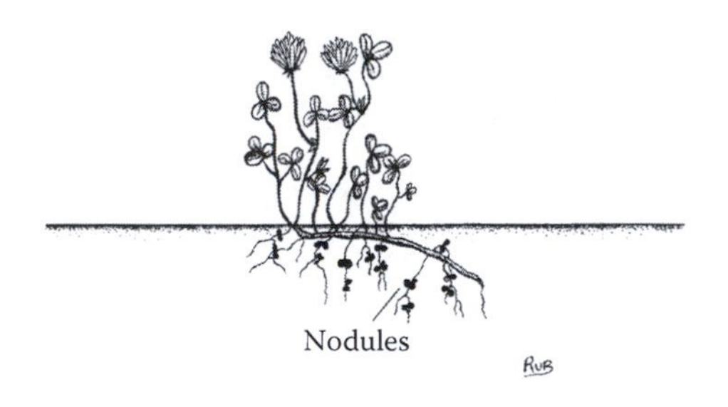

FIGURE 6.1 Legume with nodules in which nitrogen fixation occurs.

ammonium nitrogen. Legumes are members of the family Leguminosae (also named Fabaceae) and have the capacity in symbiosis with certain soil microorganisms to convert gaseous nitrogen from the air into nitrogenous chemical compounds derived from the ammonium that is fixed. The nitrogen is fixed in *nodules* in the roots and is used by the legume for root growth and shoot growth (Figure 6.1). Most of the fixed nitrogen, about two-thirds, however, ends up in the shoots. If the shoots are turned under, rotting of the organic matter releases the fixed nitrogen by *mineralization*. The amount of nitrogen added varies with the kind of legume, depending largely on the amount of shoot dry matter that is produced by the crop (Table 6.1). To obtain the yields of nitrogen noted in Table 6.1, it may be necessary to let the legume grow on the land for at least one year; otherwise, the fixation of nitrogen will be much lower than that noted, perhaps half or less. After the green manure is turned under, the benefits from the added nitrogen parallel those derived from the addition of organic matter; that is, most of the benefits are in the first season of cropping after the green manure is turned under.

Harvesting of the green manure crop by removal of the topgrowth as hay eliminates most of the benefits that would be gained from turning the crop under. In general, the amount of nitrogen removed by harvesting of the crop is equal to that added by nitrogen fixation, and the nitrogen that remains with the roots and stubble is equal to that which the crop withdrew from the soil.

TABLE 6.1
Nitrogen Fixation by Some Leguminous Crops

Crop	Nitrogen Fixed, lb/acre
Sweet clover	150–200
Alfalfa	100–150
Lespedeza	75–100
Soybeans	75–100
Red clover	65–90
Garden beans and peas	30–50

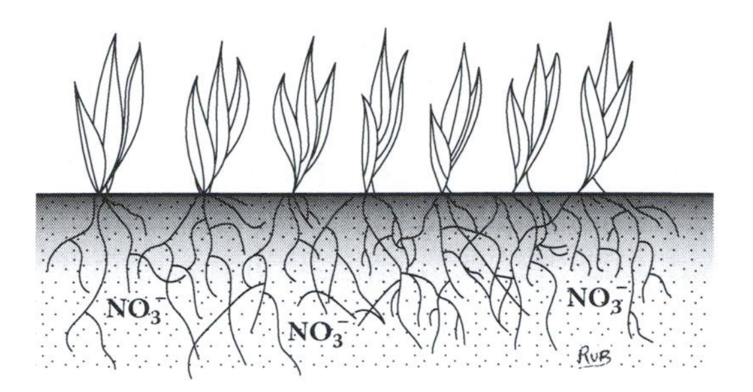

FIGURE 6.2 Fibrous root system of a catch crop, which absorbs and lessens leaching of nutrients.

Catch Crops

Nonlegumes that are grown as green manures do not add nitrogen to the soil, because they do not have the capacity to fix nitrogen. Nonlegumes, however, or legumes will help to conserve nutrients. The roots of these crops will intercept nutrients that might otherwise be lost by leaching (Figure 6.2). These green manures are called *catch crops.* Catch crops can be legumes or nonlegumes. Grasses because of their shallow, fibrous root systems are excellent for this function. Catch crops are grown after the main crops have been harvested and are turned under at the beginning of the following growing season.

Frequently in their quest for maximum yields, some growers fertilize crops in excess of the amounts that crops remove in the growing season. Some nutrients are left in the ground after harvest, and these nutrients can be lost unless a crop is planted soon after the harvest. A catch crop will utilize these nutrients and protect against their loss and potential entry into groundwater or surface water and preserve the nutrients for the following cash crop. Nitrates are likely to be lost by leaching in all soils, and leaching of potassium and phosphorus may occur in sandy soils. Some green manure crops, not grown as catch or cover crops, are reported to accumulate nutrients from deep in the soil, and when these crops are turned under, an enrichment of the surface layers of soil with plant nutrients may occur. This benefit usually is of small magnitude relative to the nutritional needs of crops.

Cover Crops

Green manures help to stabilize soil against erosion. Crops, generally nonlegumes, grown to lessen leaching or to stabilize soil are called *cover crops.* Annual grasses and small grains (particularly rye) are grown as winter cover crops. Vegetable crops usually leave little residues on the ground after harvest and leave the ground unprotected against erosion. A cover crop often is essential after these crops to protect the soil from erosion by fall, winter, and spring rains and wind. After harvest of grain crops—corn, soybeans, wheat, oats, barley, and rye—the crops residues may be sufficient in themselves to protect soil against erosion. Little protection or even losses of soil may occur if cover crops are grown after grain crops, because the ground may be bare after turning under of the crop residues and until the cover crop is established. The principal benefit of cover crops after grain crops is in the conservation of nutrients that might be leached in the absence of the cover crop. Following vegetable crop production, very little residues are left on the soil surface, and cover cropping usually is beneficial.

Selection of a Crop for Green Manuring

An ideal green manure has several desirable characteristics (Table 6.2), regardless of the function for which it is grown. A green manure crop should be easy to establish and to grow in poor soil. Infertile soils are the ones that will benefit most the green manure crop, but the primary need for good growth in poor soils is so that the farmer does not have to spend money on fertilization of the green manure. Green manures enhance the water-holding and nutrient-retention capacities of sandy soils. Clayey soils will benefit from the aggregation and improvement in soil structure that will be promoted by the additions of organic matter. Soils infertile in nitrogen will benefit from the additions of nitrogen by leguminous green manures. The green manure crop, however, should require a low level of fertility with respect to plant nutrients. To keep the cost of production low, one does not want to fertilize the green manure crop to get it to grow. However, if the soil is so infertile that the green manure crop will not grow, fertilizers should be applied. Otherwise, growth of the green manure will be so poor that no benefits of any kind will occur.

The crop should be rapid growing so that it will fit conveniently into a crop rotation. No more than one year should be required to produce the green manure crop. Any longer period of time will take the land from production of the main crops for too long. An ideal situation occurs when the supplemental crop can be grown between the time that one main or cash crop is harvested in the fall and the next one is planted in the spring. This situation occurs most often in warm climates, such as those in the southern United States. The green manure crop should cover the ground quickly to protect the soil from erosion. The total growth should be at least 18 to 24 inches to produce enough organic matter to improve the fertility of the soil.

Almost any plant that meets these criteria is suitable as a green manure crop. Some plants that are used commonly as green manures are listed (Table 6.3). Most of these crops are adaptable as green manures in a wide range of climatic conditions. Crimson clover, some vetches, and lespedeza may be suitable only in warm climates and should be treated as annuals. Hairy and smooth vetches are adapted to most regions.

TABLE 6.2
Characteristics of an Ideal Green Manure Crop

Characteristics
Is cheap to seed
Is established easily
Grows well in poor soil
Covers the ground quickly
Grows rapidly
Produces abundant and succulent top growth
Adds nitrogen to the soil

TABLE 6.3
Some Plants That Are Suitable as Green Manure Crops

Nonlegumes		Legumes	
Rye	Sudangrass	Alfalfa	Vetches
Oats	Mustard	Sweet clover	Soybean
Millet	Rape	Ladino clover	Cowpea
Buckwheat	Weeds	White clover	Lespedeza
Bromegrass		Alsike clover	
Ryegrass		Crimson clover	

If most of these characteristics (Table 6.2) are equal between a legume and a nonlegume, the legume is the better choice. A legume will add nitrogen to the soil by fixation. Nitrogen fixation is accomplished by legumes and special species of bacteria living in *symbiosis*. The bacteria infect the roots of legumes and form structures called nodules (see Figure 6.1). The bacteria are specific for a particular legume or group of legumes. If the legumes have not been grown in a soil for several years, the bacteria may not be present. Commonly, inoculum consisting of spores of the bacteria mixed with peat is purchased. The inoculum and seeds are mixed to coat the seeds with bacterial spores before planting. After the seeds germinate, the bacteria infect the root, and the symbiotic relationship is formed. The plant benefits from the nitrogen that is fixed, and the bacteria benefit from the carbohydrates that are provided by the legume. Legumes will not fix nitrogen if they are fertilized abundantly with nitrogen fertilizers, organic or otherwise. The legume will use the inorganic nitrogen derived from these fertilizers, and the nodules will not be active.

Generally legumes are more expensive to seed, harder to establish, grow more slowly, and require better soils than nonlegumes. The acidity of the soil for legume production should be between pH 6 and 6.8. Because of their slow growth, legumes might require more than 1 year in a crop rotation. Also, a good crop of a legume may be more valuable for hay on large acreage, and it would be poor management to turn it into the ground. Hay may be harvested for a year or two, and then the green manure can be plowed under to give the full benefits of a green manure crop. If a leguminous crop is harvested and not plowed under, no net increase in nitrogen in the soil occurs. A leguminous crop obtains about one-third of its nitrogen from the soil and two-thirds from nitrogen fixation. The crop has about two-thirds or more of its nitrogen in its shoots, so if the shoots are harvested, the amount of nitrogen remaining in the roots equals or is less than that removed from the soil. Growers should be aware that most of the nitrogen fixed in the roots does not remain there but is transported to the shoots.

A mixture of legumes and grasses makes a good green manure crop. The mixture ensures some success in establishing a stand. The grass will help to support the legume and may provide some winter protection. Crops, usually annuals, grown to protect another crop in this manner are called *nurse crops*. Good combinations for green manures are vetch and rye or alfalfa and bromegrass, although any successful combination for the grower's climate and soil will be suitable. The resulting green manure will reflect the high organic matter production by shoots and fibrous roots of the grass and the addition of nitrogen from the legume. Neither organic matter addition nor nitrogen addition may be as great as that which may occur in pure stands of grasses or legumes, but the compromise may be more beneficial in total addition than that which would occur from the individual crops. Grass-legume mixtures should not be fertilized with nitrogen fertilizers as the fertilizer will enhance the growth of the grass more than that of the legume.

Incorporating the Green Manure into the Soil

The green manure should be turned into the soil when the combination of growth and succulence of the crop is maximum. This condition generally occurs when the crop is at its half-height. Knowledge of the growth habit of the crop is needed to recognize this stage of growth. Another and probably more useful guideline is to turn the crop under at the initiation of bloom. Before blooming, insufficient crop growth may have occurred for enough organic matter production. After blooming, the crop begins to transport materials to the fruits and seeds at the expense of the vegetation of the plants. The C:N ratio of the vegetation will widen, and the organic matter will not be as suitable to incorporate into the soil as it was before the crop bloomed. Also, after blooming, the flowers begin to form seeds. If the seeds mature sufficiently before the crop is turned under, seeds of the green manure may germinate and become weeds in the succeeding main crop.

The amount of seed needed, the time of seeding, and the time of incorporation of green manures into the soil vary with the kind of crop and the climate (Table 6.4). Cover crops may be planted after crops are harvested. Dates in October or early November are about the maximum late dates that growers can use in northerly climates and still expect to have sufficient time to establish the crops before bad weather.

TABLE 6.4
Guidelines for Seeding and Incorporation of Green Manures

Green Manure (Kinds and Examples)	Amount of Seed		Time	
	lb/1000 sq ft	lb/acre	For Seeding	For Plowing Under
Small-seeded legumes, perennial (alfalfa, clovers)	0.4	15	Fall	Fall or spring
Small-seeded legumes, annual (crimson clover, lespedeza)	0.4	15	Spring	Fall
Medium-seeded legumes (hairy or smooth vetch)	0.8	30	Fall	Fall or spring
Large-seeded legumes (soybean)	2.5	90	Spring	Late summer
Nonlegumes (rye, oats, buckwheat)	2.5	90	Fall	Spring
Mustards (rape)	0.5	20	Fall	Spring

These dates are likely too late for the establishment of legumes. Undersowing of the cover crop in the main crop as the main crop nears maturity is common in all areas of the country to allow for establishment of the cover crop before it is too late in the growing season. This practice may not be convenient in large-scale applications without proper equipment to place the seeds of the cover crop under the canopy of the main crop.

Problems with Green Manures

A major problem with green manures is that their production takes land from the production of a main crop. The green manure should occupy the land for no more than one year in three. Unless the green manure is a valuable hay or forage crop, any more time devoted to its production may cause the costs from loss of productivity to outweigh the benefits of improved soil condition imparted by the green manure. One should use a rapid-growing green manure crop to avoid this problem or should consider using manures and composts to condition the soil. Manures and composts should be used to maintain soil organic matter on small plots of land, because area of land is at a premium and should be used for production of a main crop.

Crops that have large amounts of after-harvest residues may increase the organic matter as effectively as a green manure crop. Fertilization with some additional nitrogen to avoid immobilization is sometimes a good practice when residues of these crops are turned under in the spring. A substantial part of the soil-available nitrogen is used in rotting of crop residues in the spring. Sometimes, leaving these residues on top of the ground after harvest is more effective than trying to establish a cover crop. More soil erosion may occur between the time when the residue is turned under and when the cover crop is established than would occur if the residue were left on the soil surface until the soil is tilled for the next main crop.

Green manures should be used with caution or avoided in dry climates or on dry soils. Precipitation should exceed 18 inches per year before one considers using a green manure crop. Transpiration by the green manure will deplete the soil of water.

Deep-rooted crops, such as some of the perennial legumes, will desiccate the soil to an extent that several years may be needed to recharge the soil with water.

The green manure should be tilled into the ground at the proper stage of maturity. Turning under a mature crop may fill the soil with seeds that will produce weeds in the following season. Also, as the green manure crop matures, its C:N ratio widens. Turning under of a mature crop may cause immobilization of nitrogen and result in nitrogen deficiency in the succeeding main crop. Many growers fertilize the soil with additional nitrogen after a nonleguminous green manure crop is turned under. This added nitrogen ensures that nitrogen deficiency will not occur and accelerates the rate of decay of the organic matter so that the benefits of improved soil structure are achieved rapidly. About 50 lb N/acre (1.25 lb/1000 sq ft) are recommended to accelerate the decomposition of nonleguminous, green manures.

Some time should be allowed between working the green manure into the ground and planting of the main crops. A week or two is about the right length of time. Some green manures, such as rye, have suppressive effects, called *allelopathy*, on growth of other plants. Although this effect may be beneficial in weed control, it may delay the establishment of the main crop or stunt the crop. Also, maturing rye and other grasses may have coarse residues that may initially immobilize nitrogen or impart poor physical qualities to seed beds. Turning the grasses under at a relatively immature stage of growth helps overcome these difficulties. Some growers use wheat or oats as a winter cover in preference to rye because of the lesser growth produced by these crops. The winter-killed mat from oats is easily incorporated into the soil. Also, the mat of the oats may be planted through and left as a mulch on the surface to enable an organic farmer to produce crops without tillage or with minimum tillage. Leguminous crops with rather high nitrogen contents may need some time to decompose so that the potential injury from the initial release of ammonium nitrogen is dissipated by nitrification.

Some reports indicate that cover crops may harbor pests. Often, turning the crop under alleviates the problem, but in some cases, the pests may migrate to another crop as the crops are turned under. However, cover crops can be places where beneficial insects can harbor and help in pest control in the main crop, if the cover crop and main crop are grown on adjacent plots of land.

7 Liming

The development of soil acidity is a naturally occurring process in humid temperate climates. In humid regions where precipitation is high enough, soils become acid because of the leaching of calcium, magnesium, potassium, and other exchangeable bases from the surface layers. This leaching leaves behind exchangeable hydrogen and aluminum, which are acids in mineral soils. In cultivated land, acidity also develops from the oxidation of complex organic matter into organic acids. Some fertilizers, particularly ammonium-containing chemical or organic fertilizers and carbon-based organic fertilizers, lead to acidification of soils. Cation absorption (Ca^{2+}, Mg^{2+}, K^{+}, NH_4^{+}) in excess of anion absorption (NO_3^{-}, SO_4^{2-}, $H_2PO_4^{-}$) by plants also leads to a net production of acidity by the roots. Except in regions of relatively low precipitation, where leaching of soils does not occur and sometimes where a net movement of bases is upward rather than downward, acidification of soils is a natural process or is a process associated with crop production.

Cultivated plants differ widely in growth responses to acid soils. Crops are classified roughly into categories of acid-tolerant and acid-sensitive species. Generally, crops that have origins in the tropical or humid temperate regions of the western hemisphere are acid-tolerant because they have become adapted to the acid soils of this region. Corn, tomatoes, potatoes, garden beans, and pumpkin are acid-tolerant crops with origins in the western hemisphere. Conversely, crops that originate from the neutral or alkaline soils of the Mediterranean region, western and southern Europe, Asia Minor, and northern Africa are likely to be acid-sensitive. Many of our commonly grown vegetable crops, herbs, and flowering bulbs have been domesticated from neutral or alkaline soils of these regions.

Even the acid-tolerant plants do better in soils that are in the range of pH 6 to 7 than in acid soils, because of the favorable effects that this range of acidity has on the availability of plant nutrients.

In the United States, soils that are east of a line that extends southward from the eastern boundary of North Dakota to east central Texas will require liming, because precipitation in this region is sufficient to leach bases downward. Soils that are west of this line, with exceptions of some soils of the west coast, usually do not require liming. This north-south geographic line often is called the *lime line*.

Acid soil infertility is a syndrome of problems that affect plant growth in acid soils. This complex of problems arises from toxicities and deficiencies that occur in acid soil. Hydrogen ion concentration or acidity as such is not a major factor in poor plant growth in acid soils. A soil must be about pH 4 or lower for hydrogen ions to be toxic to plants. Among the toxic factors in acid soils are aluminum and manganese ions that increase in concentrations in the soil solution. Plant roots particularly are sensitive to injury from elevated concentrations of soluble aluminum and manganese. Seedlings and young plants are much more sensitive to soil acidity than older plants. In acid mineral soils, the detrimental effects of soluble aluminum and manganese are greater on plant growth than the effects of hydrogen ions, even at pH 4.

In certain soils with very low cation exchange capacities, calcium and magnesium deficiencies may develop, but in general, management practices such as adding certain fertilizer salts, manures, or composts adequately supply these nutrients to crops so that even in acid soils these elements are seldom deficient. However, acidification lowers the availability of plant nutrients in soil. This effect is pronounced in coarse-textured soils, which are inherently low in nutrients and which are weakly buffered against the development of acidity. Soil acidification may lead to dissolution of nutrients and to their subsequent leaching from the root zone. This effect becomes pronounced with time. In strongly acid soils ($pH < 5$), microbiological activities related to nitrogen transformations (mainly nitrification) may be inhibited, but the agricultural importance of these effects is unclear. In acid soils, the availability of phosphorus is limited by the formation of insoluble iron and aluminum phosphates. This reaction is *phosphorus fixation* (see Chapter 3), which is of major concern in growth and fertilization of crops in acid soils. Also, molybdenum becomes insoluble in acid soils to the point that some crops become deficient in this micronutrient.

Organic farmers observe that close monitoring of soil pH is important to the success of their operations. For example, rock phosphate provides little phosphorus to crops in soils above pH 5.5 or in soils below pH 5 with high amounts of iron and aluminum.

Correction of soil acidity by application of agricultural liming materials (referred to commonly as *lime*) usually alleviates the problems that result from acid soil infertility. Agricultural liming materials contain compounds that are carbonates, hydroxides, or oxides of calcium and magnesium (Table 7.1). The anions of these compounds neutralize hydrogen ions in soil solution and from the exchange sites of soil colloids and remove toxic materials from solution by precipitation.

TABLE 7.1
Liming Materials for Agricultural Soils

Material and Formula	Calcium Carbonate Equivalency
Agricultural limestones*	
Calcite ($CaCO_3$)	100
Dolomite ($CaCO_3 \cdot MgCO_3$)	108
Quicklime (CaO)	179
Hydrated lime ($Ca(OH)_2$)	133
Wood ashes (mostly calcium, potassium, and magnesium oxides and carbonates)	50

* Intergrades of limestones are mixtures of calcite and dolomite and are the most common forms of limestones in the marketplace.

In summary, there are three primary benefits of liming of soils: (1) Liming lessens the toxicities of aluminum and manganese ions by precipitating these ions from solution and lessens the toxicity of hydrogen ions in the soil solution by their neutralization. (2) The availability of soil-borne nutrients (N, P, Mo) are increased by effects that liming has on mineralization of organic matter, nitrification, and on solubility of phosphorus and molybdenum. (3) Limes are fertilizers in that they supply calcium and magnesium to soils.

LIMING MATERIALS (LIMES)

AGRICULTURAL LIMESTONES

Limestone is the principal liming material used. It is usually a variable mixture of calcium carbonate (calcite) and magnesium carbonate (dolomite), so the actual *calcium carbonate equivalency* (expression of potency relative to pure calcium carbonate; see Table 7.1) of a given limestone lies between that of calcite and dolomite. The various mixtures of limestones between calcite and dolomite are called *intergrades*. Specific intergrades are identified by their magnesium concentrations, which range from 0.6% or lower in *high-calcium limestone* (mostly $CaCO_3$) to about 13% in *dolomite* ($CaCO_3 \cdot MgCO_3$). An intergrade known as *high magnesium limestone* can have magnesium concentrations as high as 6.5%.

Calcium carbonate equivalency is an expression of the total acid-neutralizing capacity of limes in units of a weight percent of calcium carbonate (Table 7.1). The calcium carbonate equivalency of dolomite is slightly higher than that of calcite because of the lower molecular weight of magnesium carbonate relative to calcium carbonate. *Reactivity*, the rate at which limestone reacts in the soil, is a more important factor in assessing the value of limestones than calcium carbonate equivalency. Reactivity is governed by hardness, composition, and particle size. Hard materials (such as dolomite) are less reactive than soft materials (such as calcite). Finely ground materials are more reactive than coarse materials. Since increases in crop yields have been correlated with quick neutralization of soil acidity, it is desirable to use limestone that reacts rapidly with the soil.

The reaction of limestone with acid in the soil solution, using calcite as a model, is summarized by the following reaction (Equation 7.1).

$$\underset{\text{Calcite}}{CaCO_3} + \underset{\text{Hydrogen ions (acid)}}{2H^+} \rightarrow \underset{\text{Calcium ions}}{Ca^{2+}} + \underset{\text{Water}}{H_2O} + \underset{\text{Carbon dioxide}}{CO_2} \tag{7.1}$$

Neutralization is a function of the reaction of acid with the carbonate of the limestone. The hydrogen (acidity) reacts with the carbonate to form water and carbon dioxide. The calcium (or magnesium) of limestone does not bring about the neutralization, but these ions will replace hydrogen ions from soil colloids forcing these hydrogen ions into the soil solution where they will react with the limestone and be neutralized (Equation 7.2).

$$H^+ \text{ on clay or organic matter colloid} + Ca^{2+} \rightarrow Ca^{2+} \text{ on clay or organic matter colloid} + H^+ \quad (7.2)$$

Dolomite is a harder substance and is slightly less reactive than calcite. In the short run, perhaps in the year of application, equally sized particles of dolomite would not be as effective in neutralizing acidity as those of calcite. This difference will be small, especially since dolomite has a higher calcium carbonate equivalency than calcite. After the first year, differences between limestones of different composition would not be noticeable.

Generally, the reaction of limestone with the soil is complete in 2 or 3 years, but much of the reaction occurs within the first few weeks or months. This result means that much benefit of liming is achieved shortly after applications of limes to land. Finely ground limestones neutralize soil acidity more rapidly and enhance crop yields more than coarse limestones. Pulverization adds to the cost of the product, and benefits of fineness and cost of pulverization must be balanced. Mesh of limestone refers to the particle size determined by grinding the rock through a screen. Mesh refers to the number of openings per linear inch of the screen. Generally, limestone ground to pass a 60-mesh screen provides the benefits of quick neutralization of soil acidity and cost-effectiveness. Limestones ground more finely than 60 mesh to 100 mesh may be too expensive, and losses may occur through blowing of the limestone during spreading. Particles larger than 10 mesh have virtually no effect on soil pH in the short run and at normal amounts of application. Although coarsely ground materials last for a long time in the soil, they give few benefits for crop growth unless applied in amounts far greater than those used with finely ground limestones. Thus, the cost effectiveness of buying coarse limestone is lost due to the need to purchase more of that material than of the finely ground limestones.

The difference in neutralizing capacities between calcite and dolomite increases as the particles become coarser, but with particles of 60 mesh or finer, the differences are very small. In practice, most agricultural limestones are mixtures of calcite and dolomite so that if the particles are finer than 60 mesh, differences among types (intergrades) of limestones may be undetectable. Dolomitic limestones have the advantage of providing magnesium as well as calcium and would be the preferred choices on this basis. Analyses for calcium and magnesium are reported on the bags of limestone and should be available from the vendor of bulk-spread limestones. Repeated use of high-magnesium limestones or dolomite should be monitored by soil testing to ensure that magnesium in the soil does not reach concentrations that might limit crop production.

Quicklime and Hydrated Lime

Quicklime is sometimes called *burnt lime* or *garden lime*. The heat from burning materials decomposes the carbonates of limestone to oxides with loss of carbon dioxide to the air. Calcium and magnesium oxides are formed in proportion to the amount of calcium and magnesium carbonate present. The following equation demonstrates the reaction that occurs with calcite (Equation 7.3).

$$\underset{\substack{\text{Calcium carbonate} \\ \text{(Calcite)}}}{CaCO_3} + \text{Heat} \rightarrow \underset{\substack{\text{Calcium oxide} \\ \text{(Quicklime)}}}{CaO} + CO_2 \quad (7.3)$$

For rapid neutralization of pH or to raise the pH higher than 6.5, growers should consider using quicklime or hydrated lime. Hydrated lime is formed by reacting quicklime with water (Equation 7.4).

$$\underset{\substack{\text{Calcium oxide} \\ \text{(Quicklime)}}}{CaO} + \underset{\text{Water}}{H_2O} \rightarrow \underset{\substack{\text{Calcium hydroxide} \\ \text{(Hydrated lime)}}}{Ca(OH)_2} \quad (7.4)$$

This process is called *slaking*; hence, the name *slaked lime* is used frequently for hydrated lime. Hydrated lime mixed with water is *whitewash*. Quicklime and hydrated lime are not considered to be organic limes by certifying organizations.

Basic reactions for the neutralization of acidity by these compounds are as follows (Equations 7.5 and 7.6).

Quicklime

$$CaO + 2H^+ \rightarrow Ca^{2+} + H_2O \quad (7.5)$$

Hydrated lime

$$Ca(OH)_2 + 2H^+ \rightarrow Ca^{2+} + H_2O \quad (7.6)$$

The neutralization occurs by the reaction of the hydrogen ions (acid) with the oxide or hydroxide and the resulting formation of water. The calcium ions that are formed will replace adsorbed hydrogen ions on the colloids, forcing them into solution where they will be neutralized by the limes (Equation 7.2).

Before 1900, much of the lime used was quicklime. Burning served at least two purposes other than providing quick reactivity of the lime with soil. Burning reduced the weight for hauling (loss of carbon dioxide gave a 44% loss of weight), and since rock or shell crushers were not conveniently located, burning facilitated reducing the product to the proper state of subdivision, as the burnt lime is more amorphous than the crystalline limestone or shells. Quicklime is not used often today, except when the grower has a need to accelerate the raising of soil pH. Organic certifying agencies do not certify quicklime for organic agriculture, because the agencies feel that quicklime is too reactive for safe use with crops.

Wood Ashes

Burning of wood, bark, or paper products leaves ashes, the composition of which depend on the amounts of alkaline metals (calcium, magnesium, potassium) present in the original materials. The first products are oxides. With aging, these oxides will be slaked to hydroxides and ultimately will react with carbon dioxide in the air to form carbonates. The effectiveness of wood ashes as lime depends on their composition, which varies with the product burned and with handling (aging, leaching). In general, wood ashes have about half or more of the calcium carbonate equivalency of agricultural limestone. Ashes of hardwoods are likely to have a higher calcium carbonate equivalency than ashes of softwoods because the hardwoods have higher concentrations of calcium, magnesium, and potassium than softwoods. Wood ashes add considerable potassium to the soil and are considered to be a potassium-containing fertilizer. They also add substantial amounts of calcium and magnesium to the soil. Wood ashes do not add much phosphorus.

Management Related to Application of Limes

Response to limes varies for different crops and even for varieties of crops on the same soils, and this response can be determined only through experimentation. However, most growers are not interested in laying out a liming experiment to determine the lime requirements of their soils. Hence, soil testing is a common means for estimating lime requirements. Soil pH is the principal criterion used in determining the need for lime. Soil pH is an expression of *active acidity*, which expresses the hydrogen ions in the soil solution (Figure 7.1). However, most of the acidity of soils resides on the soil colloids, which are primarily clays and organic matter. This fraction, referred to as *reserve acidity*, constitutes the major portion of soil acidity. Only a small amount of limestone, a few pounds, may be needed to neutralize the active acidity, whereas thousands of pounds are required to neutralize the reserve fraction. Because of this fact, lime requirement varies with soil texture and pH. A fine-textured soil (e.g., loam, silt, clay) will require a larger amount of liming materials to reach a desired pH than a coarse-textured soil (e.g., sand, loamy sand, sandy loam). Some guidelines for determining the lime requirement of soils of various textures are given below (Table 7.2).

Examples of application of the above guidelines (Table 7.2) are as follows. A sand would require 1.5 tons of limestone to raise it from pH 5 to pH 6.5 (1.5 pH units times 1.00 ton per unit). A sandy loam of pH 5 would require 2.0 tons of agricultural

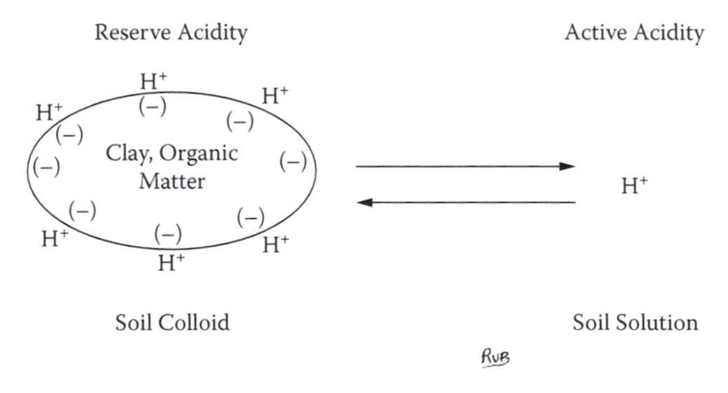

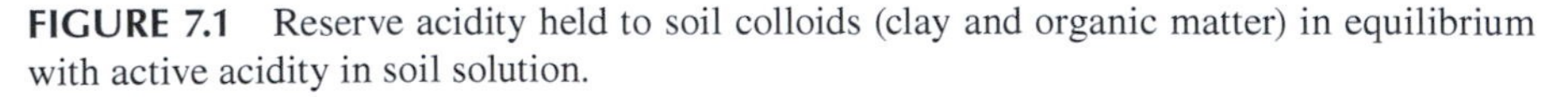
FIGURE 7.1 Reserve acidity held to soil colloids (clay and organic matter) in equilibrium with active acidity in soil solution.

TABLE 7.2
Agricultural Limestone Needed to Raise pH of Soil by One Unit*

Soil Textural Class	Lime Requirement (Tons/Acre)
Sand	1.00
Sandy loam	1.33
Loam	2.00
Silt and silt loam	2.33
Clay and clay loam	2.50

* Recommendations are based on 60-mesh agricultural limestone incorporated 3 inches deep in the soil.

limestone to raise it to pH 6.5 (1.5 pH units times 1.33 tons per unit). A clay loam for the same adjustment in pH would require 3.75 tons of limestone (1.5 pH units 2.50 tons per unit). The more potent materials, quicklime and hydrated lime, would be used in lesser amounts than agricultural limestone. Use about 50% as much quicklime (calcium carbonate equivalency of about 180) and about 75% as much hydrated lime (calcium carbonate equivalency of 133) to bring about the same adjustment in pH as that given by addition of agricultural limestone. To lime the sandy loam in the previous example to pH 6.5, 1.1 tons of quicklime (rather than 2 tons of agricultural limestone) would be used (Table 7.3). The amount of hydrated lime needed for the same adjustment would be 1.5 tons per acre. Some other conversions are presented in Table 7.3.

Since the guidelines based on pH and soil texture are only rules of thumb, more accurate methods of estimating lime requirements have been developed by soil-testing laboratories. The results of these tests are reported as *buffer pH* or the *lime requirement* (or as lime index, the amount of lime needed to raise the soil to a target level of pH 6.5 or so). The buffer pH is an expression of the sum of active and reserve acidities, or what is sometimes referred to as titratable acidity and is used in the soil-testing laboratory to determine the lime requirement from experimentally derived

TABLE 7.3
Examples of Amounts of Various Limes Needed to Raise a Sand, a Sandy Loam, or a Clay Loam from pH 5 to pH 6.5

	Amount of Lime Needed (Tons/Acre)			
Texture of Soil Lime	Agricultural	Quicklime	Hydrated Lime	Wood Ash
Sand	1.5	0.8	1.1	3.0
Sandy loam	2.0	1.1	1.5	4.0
Clay loam	3.75	2.1	2.8	7.5

data presented in charts. Buffer pH is not informative to growers. Growers should refer to regular pH measurements to assess soil acidity.

Normally, lime should be applied as far ahead of seeding or planting of a new crop as practicable. Generally the recommended amount of time is 3 months to a year. The fall before a spring crop is planted is a convenient time for liming as this time allows for limes to react with the soil before spring planting. In the fall, also, fields are likely to be dry, making it easier to get equipment across them. Commonly, growers fail to make the application until after the seedbed is prepared in the current year. In this case, quicklime or slaked lime might be used, although finely ground agricultural limestones applied at planting usually give the desired results of raising pH sufficiently for crop production in the current year. Generally, not more than 0.5 to 0.75 ton of quicklime or hydrated lime should be applied per acre per year. These materials work very quickly in the soil. Their effects also diminish faster than those of agricultural limestone. Note that quicklime and hydrated lime are not permitted as organic materials by certifying organizations.

Agricultural limestones are applied to soil surfaces in amounts of 1 to 6 tons per acre and normally are mixed into the soil by tillage—disking or harrowing into plowed fields. Often a lack of uniform mixing in the plow layer occurs. Poor mixing of lime in the soil may lead to poor neutralization of the soil in the short run. Over a period of time and with further tillage and mixing, the neutralization becomes uniform. Uniform lateral application across the soil surface is important. This step is more important than uniform mixing vertically into the soil. Lime applied at recommended rates does not need to be mixed into the soil more than about 3 inches. It is important that the pH be adjusted correctly in this zone to protect germinating seeds from acid soil and for seedlings to become established. If the lime is mixed into the soil at greater depths, the amount of application should be increased to keep the weight of lime per unit weight of soil constant.

Plowing under of limes with a moldboard plow is not recommended, since this tillage operation may bury the lime too deeply in the soil and leave the surface zone acidic and unreacted with the lime. Seeds and seedlings in the surface zone will not get benefit from the lime that is plowed under in a zone out of reach of the plant roots. Disking or harrowing lime into a plowed field is the best practice for incorporating lime into soils.

Because of poor mixing of lime or uneven spreading across the surface, pockets of high lime concentrations may occur. This condition occurs frequently when lime is mixed with wet soils. These pockets may lead to a condition known as the *overliming effect*. Because of the possibility of the occurrence of the overliming effect, agronomists recommend that no more than 4 tons of limestone be applied per acre per year, even if the lime requirement exceeds 4 tons per acre. Lime requirements in excess of 4 tons per acre should be met with applications split between 2 years. For example, a 6-ton requirement could be met with an application of 4 tons in one year and 2 in the following year or with 3 tons in one year and 3 in the next year. In most cases the effects of overliming go unnoted because their adverse effects are less than the beneficial effects of correcting soil acidity (Figure 7.2). The causes of the detrimental effects of overliming are not known, but they are short-lived and usually are not apparent after 2 years, at which time the maximum benefits of liming

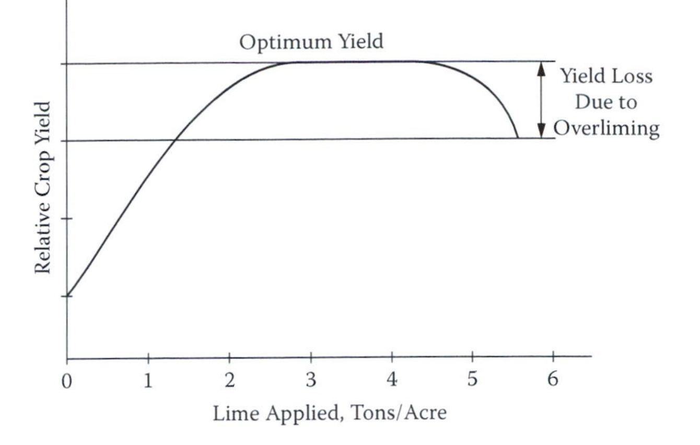

FIGURE 7.2 Diagram of crop yields as a function of limestone application in one year. The yield suppression is due to overliming that occurs in the first year or two following applications.

will be expressed in crop yields. The principal problem of poor mixing of lime with soils is the incomplete neutralization of acidity and resultant slow change in pH. Overestimations of the lime requirement and dumping of wood ashes on land are common errors that lead to overliming.

Once a plot has been raised to about pH 6.5, it will remain at a constant level with moderate applications (for example, 2 tons/acre) of lime at intervals of about 5 years. More frequent applications of lime will keep the pH more constant from year to year than liming at intervals of 5 years or more. Needs for reliming are governed by soil texture, leaching, cropping, cultivation practices, fertilization, and nature and fineness of the original liming materials, as well as by the crop to be grown. The pH should be checked every 2 to 3 years to ensure that the acidity has not increased into a range that would restrict crop yields. It is recommended commonly that soil pH not fall below 6.0.

8 Mulches

A mulch is a covering layer of material applied over the surface of land. The layer of mulch creates a buffer between the soil and aerial environment. Mulches are used for many different functions and may be composed of many different materials (Table 8.1). The functions of mulches include water conservation, weed control, temperature regulation, erosion control, and ornament, among other things. Commonly recognized mulching materials include straw, hay, composts, sawdust, wood chips, and bark. Plastics of varying colors also are well-known agricultural mulches. Lesser-known materials are paint, gravel, stones, aluminum foil, paper, and carpeting. An organic mulch is a layer of organic matter covering the surface of the land.

FUNCTIONS OF MULCHES

Water Conservation

This practice is perhaps the most common reason for applying mulches. A mulch for this purpose should decrease the evaporation of water from soil and should increase the infiltration of water into soil (Figure 8.1). Water rises naturally in the small pores and channels of soils through capillary action. A mulch breaks up the capillary rise at the soil surface and inhibits evaporation of water into the atmosphere. Also, the soil is protected from drying by direct sunlight and wind. Mulches decrease runoff over the soil surface and enhance penetration of water into the soil. On a bare soil, precipitation may not enter the soil as easily as it will run across the surface and off the land. Mulches absorb the energy of rainfall, impede runoff, and increase the time that water is in contact with a given area of soil, thereby increasing infiltration of water into the soil.

Mulches for water conservation should be loose materials that disrupt the upward flow of water, maintain a high humidity within the mulch, and allow water to percolate freely through the covering. These mulches should be 2 to 4 inches deep, being a little deeper on sloping land than on flat land. Thin mulches shift and decay and may not protect the soil with continuous cover over a growing season. The mulch should not be a water-absorbent material and should allow water to pass through it and into the soil.

A water-laden mulch is of no value to a crop and possibly may be detrimental because the wet mulch might be a haven for plant diseases. Extremely deep mulches, more than 6 inches deep, may hold water and prevent its percolation into soil. Plants that are rooted in the soil, therefore, will be deprived of water rather than having their moisture status improved by the mulch. Another problem with deep mulches is that roots will grow in the mulch and not develop in the soil. This condition can lead to the plants obtaining insufficient nutrition and water.

TABLE 8.1
Functions of Mulches and Suggested Materials

Function	Materials
Water conservation	Straw, hay, sawdust, woodchips, composts, manures, dust, plastic
Weed control	Straw, sawdust, woodchips, leaves, black plastic, paper, hay, composts, manures
Temperature regulation	Plastic (soil warming), organic materials (soil insulation)
Sanitation	Plastic or organic materials
Erosion protection	Organic materials, plastic, burlap, stones, gravel
Plant nutrition	Composts, manures, hulls, leaves, plastic (to prevent leaching)
Insect repulsion	Light colored mulches: white plastic, aluminum foil, foil-covered plastic or paper
Ornament	Bark, woodchips, sawdust, peat moss, hulls, sand, gravel, dust

Salt marsh hay is a highly preferred, although scarce, material for water-conserving mulches. Its stiff stems resist packing and provide an ideal cover. Other good materials for water conservation in descending order of quality are straw, hay, grass clippings, wood chips, composts, and manures. In dry climates, a dust mulch, created by tillage 1 or 2 inches deep, will impede evaporation. In humid regions, dust mulches must be created after each rainfall. Peat moss prevents evaporation from the soil, but peat moss absorbs water and does not allow precipitation to pass freely into the soil. Peat moss is difficult to wet once it dries, and on the surface, dry peat moss may inhibit percolation of water into the soil, causing the water to run off instead.

Plastic mulches restrict evaporation but also inhibit movement of water into soil. Before plastic mulch is applied, it is critical that the soil water level be at or near field capacity. Water cannot be applied effectively by rain or overhead irrigation to crops growing under plastic mulch. The plastic must be cut with slits to permit passage of

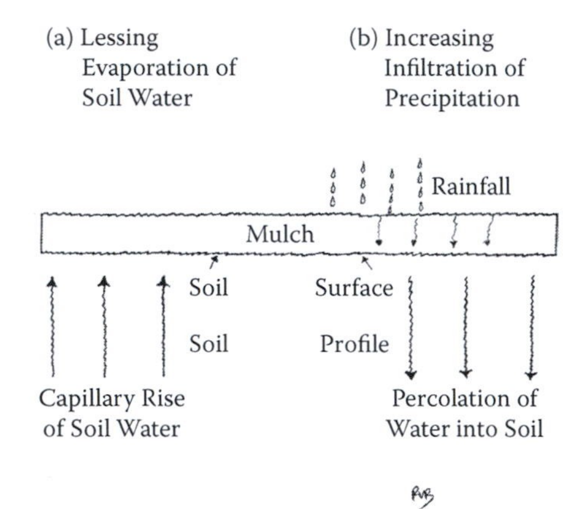

FIGURE 8.1 (a) Restriction in evaporation of water from mulched soil and (b) increased infiltration of water into mulched soil.

water. Water may be provided under plastic by drip irrigation for which the tubing for application of water must be applied before the mulch is laid.

Weed Control

In humid regions, this function may be the most important one. Mulches for weed control are particularly valuable when labor is in short supply, seasonal, or expensive. Mulching for weed control may make the difference in having a vacation or not having a vacation from the garden. Mulched crops require no tillage or at least a minimum amount of tillage for weed control (see Chapter 9). Because of shallow rooting in mulched soil, weeds are easy to pull. In addition to labor savings, mulching helps to avoid root or foliar injury that may occur by mechanical operations during cultivation of unmulched crops.

Mulches for weed control should be thick enough or impervious enough so that weeds cannot grow through them (Figure 8.2). Straw or sawdust mulches should be 2 to 4 inches thick after packing. A similar mat of hay can be used, but it may contain crop or weed seeds that will infest the soil. Maintenance of a several-inch-thick layer of hay helps to avoid the problem with weeds, but if the hay mulch decays or is turned under, weeds will be a serious problem if the soil is not mulched again. Hay for mulch should be inspected to ensure that it is free of matured seeds. Spoiled hay (hay that has been rained on) often is used for mulch. This hay might contain seeds if the hay was harvested at a mature stage (flowering) of plant development.

Composts are excellent for weed control. Properly made composts should contain few weed seeds unless weeds have grown in the compost pile, flowered, and produced seeds. Immature composts in which organic matter has not decayed sufficiently may be infested with weed seeds if the material from which the composts were made was contaminated with weed seeds. Raw manures are not as good as composts for weed control. In fact, manures likely contain weed seeds from forages consumed by livestock. Manured plots usually become weedy, and mulching must be intensified by using thick or impervious layers on manured land. If manures are used as a mulch for any function, composted manures or ones with a lot of bedding make a better surface for walking and for maintenance of crops than fresh manures with low bedding contents. Manures and composts are fertilizers, and a 2-inch thick layer may provide enough nutrition to grow a crop. The nutrients leach with rainfall or irrigation from the mulch and into the soil.

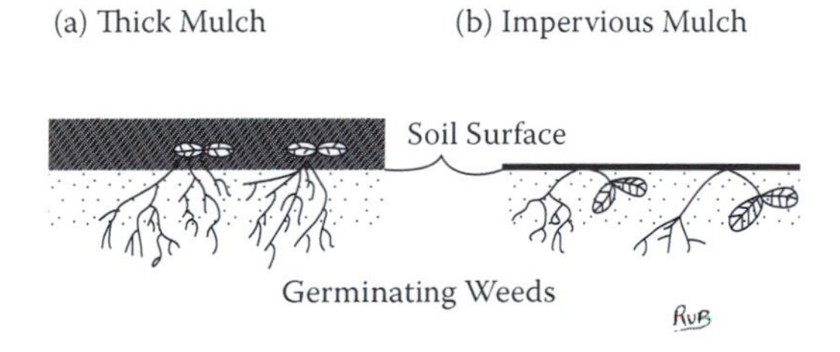

FIGURE 8.2 Weed control by (a) a mulch that is too thick for weed emergence and (b) a mulch that is too impervious for weed emergence.

Aggressive weeds, such as nutsedge, docks, perennial grasses, and Jerusalem artichoke, will grow through ordinary mulches, even through mulches that are several inches thick. Impervious materials are needed to contain these weeds. Any affordable sheets of impervious, nontoxic material, such as plastic or heavy paper, are effective. Removal of plastic at the end of the season presents a problem. Watering of crops is a problem with the plastic also. Paper is the best material to use as a barrier mulch for weed control under a layer of organic matter. A layer of newspaper about 4 to 8 sheets thick placed underneath 1 or 2 inches of straw, compost, or manure will control most weeds. At the end of the season, the newspaper will be decomposed with little trace of material requiring no removal. No soil contamination results from the use of black-and-white newspaper. Inks on these papers are carbon and vegetable oils. Colored print and magazines might contain metals that the user would want to avoid.

One also has to be careful in the use of plastic sheets under deep, permanent mulches on perennial plantings. Cases have been reported in which roots of perennials have grown into the layer of mulch over the plastic. These roots may constitute the major root system of the plants, and in dry weather they would be cut off from the soil by the plastic (Figure 8.3). Some people have almost lost their crops as a result.

Paper pulp, shredded paper, and shredded vegetable garbage harden into an impervious mat. For example, vegetable garbage can be ground in a garden shredder and poured onto the ground in a 2-inch layer, which will dry into a thin impervious mat with a texture somewhat resembling that of ceiling tiles. These layers are impervious to emerging weeds but let water penetrate into the soil.

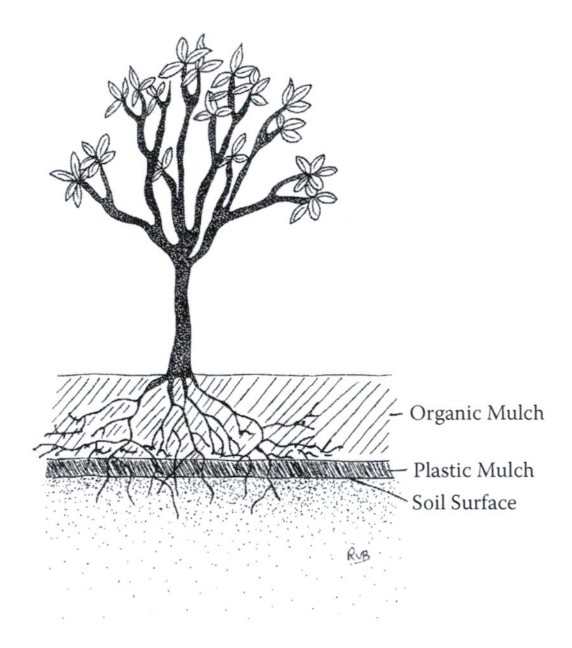

FIGURE 8.3 A perennial crop, such as blueberry, rooting into a thick layer of organic mulch placed over plastic on a soil surface.

A dust mulch has some effect in control of weeds. Seeds will not germinate in the dust mulch because it is too dry. Sand, gravel, and stones give short-term benefits in weed control. In the long term, however, weed growth in the sand or gravel and around the stones destroys the effectiveness of these materials. These weeds can be controlled only by hand or by herbicides. One often regrets putting down stones or gravel as mulch to control weeds.

Temperature Regulation

Temperatures in soils are stabilized by the insulating effects of thick layers of organic matter. Plastic layers warm soils and facilitate early planting of crops in the spring. For winter protection of plants, thick layers of organic matter are used. These materials should be applied after the ground has become cold after a few light freezes or after it has frozen. Their purpose is to keep the ground cold and to prevent the action of freezing and thawing in the soil. Freezing and thawing and the resulting heaving and subsiding movements may push plants out of the ground or snap their crowns from roots. Straw is the best readily available material for winter protection. Hay is also good. Avoid unshredded broad leaves because these may pack and inhibit emergence of shoots in the spring. Also, damp conditions under the packed leaves may promote diseases that rot crowns of perennials. Mulch only up to the crown if broad leaves are applied. Pine needles or shredded leaves can be used more successfully than unshredded broad leaves. Hardy plants that maintain some foliage over winter are often better left unmulched or should be mulched under their leaves, taking care not to cover the evergreen leaves or crown of the perennials. Mulches should not be applied all of the way up to the trunks of trees. A depression should surround the trunk so that insects and diseases that might develop in the organic mulch do not infest the trees.

It is not likely that an organic mulch can be applied deep enough to prevent the ground from freezing. In this respect, a 6-inch deep cover of snow provides more protection and insulation than a comparably thick organic mulch. The organic mulches may provide protection of shallow roots against dehydration from exposure to winter weather.

A soil covered with organic mulch will warm slowly in the spring and may delay planting of annual crops. At 4-inch depths, soils under raw leaves will be as much as 20° cooler than bare soils. The mulches should be removed from areas to be planted to annual crops. The mulches can be replaced as soon as the soil warms. Mulches left on perennial crops in the spring will delay warming of the soil and will slow plant growth in the spring. This effect is not entirely disadvantageous. An advantage of this action may be delaying early growth or flowering during unseasonable warm spells. In general, however, mulches applied for winter protection should be moved away from perennials after the danger of killing frosts has passed.

For enhanced warming of soils, use plastic mulches. Clear plastic mulches are the preferred ones for promoting earliness of crops. Clear plastics allow for greater warming of soils than black plastics. Although results vary among investigators, a clear plastic mulch may permit warming of 6 to 10°F to a depth of 6 inches, whereas black plastics permit warming of 1 or 2°F to 1 or 2 inches. Sunlight passes through the clear plastic and heats the soil. A layer of water on the underside of the plastic retains the radiant heat at night through what is known as a *greenhouse effect*. With black plastics, the mulch absorbs most of the sunlight and becomes greatly warmed, and little energy passes through to warm the soil. Shallow-rooted crops benefit from the warming permitted by black plastic and receive the added benefits of weed control and having the fruit kept off the ground as the season progresses.

On a frosty night, the air above a plastic-mulched crop will be colder than that above bare soil, and foliar damage to the crop is likely to occur. Row covers or hot caps may be necessary to protect the crop. Ordinarily, to limit the costs of using plastic mulches, only the area around the plants is covered, leaving some bare ground between rows. Some warming of the air around the crop will occur from irradiation by the bare ground. Light-colored plastic mulches, such as white plastic or aluminum-painted plastic reflect light. Soil temperatures under these mulches will be cooler than those under bare surfaces.

With perennial crops, plastic can be left on site, but with annual crops, removal of plastic mulch at the end of a growing season usually is a necessity. Most plastic mulches decompose in sunlight but do not decay in soil. Even biodegradable mulches require light for their breakdown. Problems with biodegradable mulches are that they may not last for the season on the ground, but portions under the ground will show little indication of decomposition. Plastic that is used on greenhouses is thicker (4 or 6 mil; 0.004 in. or 0.006 in.) than mulching plastic (1 or 1.5 mil) and is manufactured to be resistant to light-induced breakdown.

Experimentation with bituminous paints sprayed on the soil has shown that these materials are not effective in warming soils. Checking and cracking of the soil surface destroys these mulches, quickly dissipating the effectiveness, if any, that they may impart. Other black materials that might be used are soot, coal dust, and black paper. These black materials do not give warming equivalent to that of black plastic.

Other Uses of Mulches

Sanitation

Mulches will keep produce cleaner by keeping mud from splashing on upright or fallen plants. The mulch will lessen disease infection on foliage or fruits that come in contact with the ground. Organic mulches should not contain diseased plants or refuse from the plants that are being mulched.

Erosion Protection

On sloping sites on which ground cover has not become established after tillage, mulches can protect soil against erosion. Burlap or netting will help hold soil in place until plants become established. These materials are useful in applications in which the crop emerges through the mulch. These materials are useful in establishment of lawns or other seedings on slopes. The mulch must give no impediment to seedling emergence while protecting the soil from erosion. Straw, hay, wood chips or shavings, or plastic may be used around or between rows of newly transplanted crops to lessen erosion.

Ornament

Many mulches enhance the appearance of plots as well as provide other useful functions. Bark mulches are valued highly for their ornamental value. Wood chips because of their bright appearance when fresh are considered less desirable than bark, but they soon weather into attractive mulches. Wood chips normally are much cheaper than shredded bark or chunks of bark. Cocoa hulls, cottonseed hulls, or buckwheat hulls make attractive mulches. Gravel, crushed rock, and stones may provide walkways as well as attractive mulches. Bark, wood chips, hulls, and the gravel, rock, and stone mulches may be expensive to purchase, apply, and maintain. Expense of purchase of bark, wood chips, and hulls may prohibit their use in amounts needed to impart benefits of weed control or moisture conservation. Weed growth in gravel, rock, and stone mulches is a frequent problem.

Insect Control

Reflective mulches, such as aluminum foil, foil-coated paper, or aluminum paint applied to plastic or directly to the soil have been demonstrated to repel insects. These materials are effective in repelling aphids that carry viral diseases that infect squash. Apparently, the reflective surfaces confuse the insects so that they do not land in the plots. Reflective white plastic mulches also have this capacity.

Plant Nutrition

Manures, composts, hulls, leaves, and other plant residues contain plant nutrients, and sufficient nitrogen and potassium may leach from these mulches to fertilize a crop in a current season. Other nutrients generally are too low in concentrations in the mulch or too immobile in the mulch and soil to provide much nutrition to a crop in one season. By inducing shallow rooting, mulching promotes root growth in the top soil. This action permits plants to obtain nutrients in the richer zones of the soil and allows plants to obtain nutrients that have been applied to the soil surface and that have moved only a short distance into the ground.

Leaching Control

On sandy soils, plastic mulches have been employed to inhibit leaching of fertilizers. In this use, mulch is applied over, sometimes elevated, beds in which concentrated bands of fertilizers are placed. The water table should be high in these soils; otherwise, the soil may become droughty.

Living Mulches

Sometimes, plants (living mulches) are grown between rows of crops, sharing the same plots. Living mulches help to control soil erosion, crowd out weeds, aid in pest control, and sometimes add nitrogen. The objective is to grow the living mulch without harming the crop. Living mulches, even leguminous plants, likely provide no nutrition to crops and may compete with the crops. Generally, the main crop should grow in an 18-inch-wide strip that is free of the living mulch. Living mulches should be low-growing so as to not shade the crop. Sometimes the mulches are mowed to provide paths between rows of crops and to limit the competition of the mulches with the crop. The mulches can be quite attractive in annual or perennial plantings and provide walkways between rows of plantings. Alfalfa and red clover for northern climates and subterranean clover for southern climates are possible mulches for interplanting with row crops. Use of the subterranean clover has been successful because it dies off in warm weather and does not compete with the crop. Living mulches should be used with caution in dry soils, for the mulch may dry the soil by transpiration. Deep-rooted crops such as alfalfa could cause soils to become droughty. Aid in insect control may come about from predators that may be in the living mulches, but a potential worry about living mulches is that the mulch may attract pests that are not affected by the predators and that the pests will move from the mulch to the crop.

Problems with Mulches

Cost

Large amounts of materials are required for mulching. Polyethylene mulches cost several hundred dollars per acre (4-ft wide strips placed on 5-ft centers). A typical cost might be $1000 per acre to purchase organic materials. A bale of straw weighing about 50 lb will cover about 25 to 50 sq ft of land. At least 20 tons (800 bales) of straw are needed to mulch an acre. Mulching with manures or composts may require 100 tons of these materials per acre. If one is to grow the materials, areas of land to grow the mulches need to be at least twice the area of land to be mulched. Application of mulches is time consuming, and labor to apply mulches is expensive. On small garden-sized plots, total costs of labor and materials are low, and the convenience of having the mulches for weed control outweighs the costs.

Wrong Application

Mulches tend to maintain soil conditions in the state that existed when the mulches were applied. A cold soil will remain cold, a dry soil will remain dry, and a wet soil will remain wet. Prolonging these conditions by mulching may not favor crop growth.

Packing

This problem was mentioned previously in discussions of including broad leaves in mulches. Care should be taken when mulching perennials to ensure that their crowns are not covered by mulch.

Decomposition

A mulched crop will have a shallower root system than an unmulched crop (Figure 8.4). If the mulch decomposes, the shallow roots, which may be practically at the soil surface, will be exposed. This problem is common in woody, perennial plantings such as blueberries. Sawdust and wood chips decompose slowly and make good mulches for woody perennials. Inspections should be made at least annually to ensure that roots in perennial plantings are covered with mulch, and additional mulch should be applied as needed.

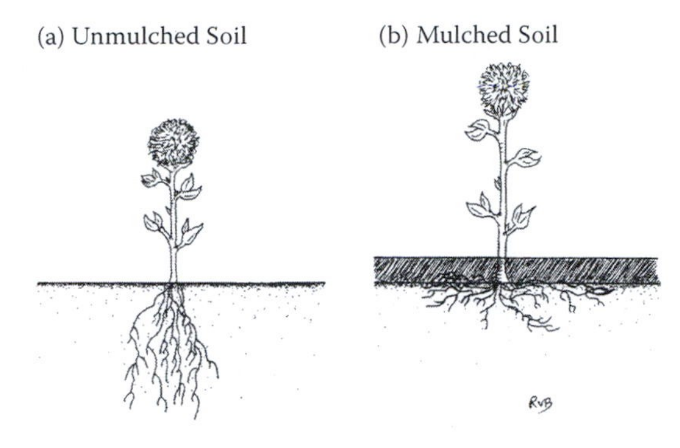

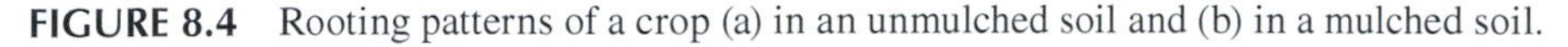
FIGURE 8.4 Rooting patterns of a crop (a) in an unmulched soil and (b) in a mulched soil.

Plant Nutrition

Manures, composts, and hulls contain plant nutrients, primarily nitrogen and potassium. These nutrients will leach from the mulch and into the soil. Ordinarily, these nutrients are beneficial, but if a crop does not need any additional nutrients, harm may be done to the crop. Ornamental mulches of cocoa hulls have been known to cause foliar burning of rhododendrons. This burning is caused by the potassium that leaches from the mulch and that is then accumulated to excess by the plant.

Bark, wood chips, sawdust, and dead leaves are so low in nutrients that toxicity from these materials is unlikely. One need not be concerned that decomposition of these mulches will affect plant nutrition by immobilization of nutrients. As long as the mulches are on top of the ground, little immobilization of nitrogen or any nutrient will occur in the soil, and any immobilization would be principally at the soil–mulch interface. If, for some reason, these mulches are mixed into the soil, nitrogen deficiency may occur and last for up to 3 years. For every 100 lb of mulch turned in, 1 lb of nitrogen should be added to the soil at least once during the first year.

Generally, one should not apply fertilizers to organic mulches as long as they are on the surface, since this practice would hasten their decomposition. However, if crops need fertilizing, fertilizers can be applied to organic mulches, and the nutrients will leach through the mulch and into the soil.

One does not need to be concerned with acidification of soil under mulches. Some acidification may occur at the soil–mulch interface from organic acids produced in the mulch, but this acidity is dissipated soon after it enters the soil. A thin dusting of lime under the mulch before its application or on the surface of the mulch after its application will neutralize any acidity that may develop. Application of lime to the surface of mulches is an acceptable practice. With plastic mulches, all of the fertilizer necessary to grow the crop should be applied in the zone to be covered with mulch or distributed with irrigation water under the mulch.

Mice and Other Varmints

Rodents often live in the mulch and feed on the mulched plants and may create problems particularly over winter. Woody plants may be girdled and killed. Plants should be protected by leaving space between them and the mulch or by wrapping them with protective coverings. Protection of plants against rodent damage is warranted particularly on snow-covered plots.

Fire

Deep, dry layers of organic material, such as hay or straw, will burn if lighted.

9 Tillage

Land is tilled to put the soil into proper physical condition for the germination of seeds and growth of crop roots, to incorporate green manures and other residues, to mix in manures, composts, and fertilizers, and to control weeds. Tillage also helps in aeration and in drying and warming of soil. Shallow tillage can create a dust mulch that aids in water conservation. Tillage has effects on control of pests other than weeds. Aeration of soil after tillage may aid in control of diseases. Tillage exposes insects to adverse conditions in which they are vulnerable to predators, freezing, and desiccation.

PREPARATION OF SEEDBEDS

As mentioned above, a principal objective of tillage is to develop good physical condition in the soil for plant growth—or good *tilth*. Soil with good tilth has a well-aggregated structure. *Aggregates* are small, preferably sand-sized, structures formed by arrangement of soil particles into groups that are held together tightly. Sand-sized aggregates formed from clay impart good tilth to this fine-textured soil. These soils will be crumbly and loose or *friable*. With sandy soils, development of good tilth is not of the concern that it is with clayey soils. Sandy soils are called *light soils* because of their ease of tillage and possess good tilth naturally. Gravel-sized and larger structures are called *clods* and impart poor tilth. Cloddy soil usually does not provide good conditions for germination of small-seeded crops. Cloddy soils result from the compression that results from tillage of wet, fine-textured soils and the failure of the particles to break into small aggregates.

Proper tillage can form aggregates, but aggregates will be formed only in soils that have clay or organic matter or both constituents. Sands and silt will not cohere to produce stable aggregates, although these particles may be included in aggregates of clay. Organic matter increases the stability of aggregates and helps in the formation of sand-sized structures with clay, thereby promoting good tilth. Growth and decay of plant roots and green manure crops promote aggregation and improve the stability of aggregates. Freezing and thawing of soil improves aggregation by breaking up clods. Fall plowing is employed often on clayey soils to develop good tilth. The freezing and thawing actions over the winter help in production of an aggregated structure. However, sloping land generally should not be plowed in the fall because of the high potential for erosion of the unprotected, bare soil.

A seedbed with good tilth is granular, well aerated, and well drained, and has good water-holding capacity. It also has a good capacity to receive water by capillary rise from depths of the soil as well as to absorb water from precipitation or irrigation. Proper tillage can help develop good soil tilth. Improper tillage can ruin tilth. Tillage of wet soil should be avoided. Compression of fine-textured, wet soil by

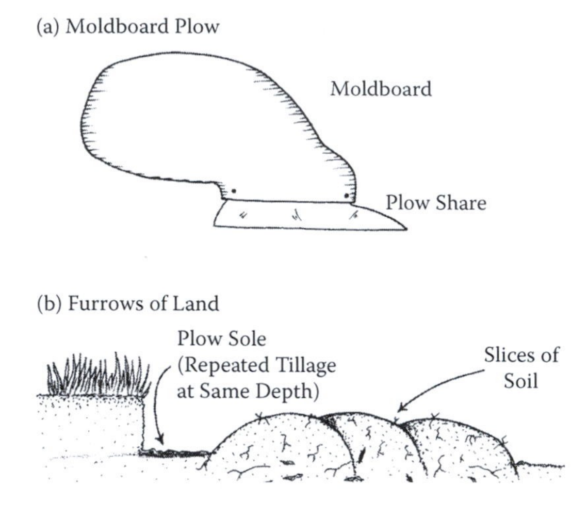

FIGURE 9.1 Diagrams of (a) a moldboard plow and (b) land tilled with a moldboard plow.

tillage implements, such as the moldboard plow (Figure 9.1) or rotary tiller, can lead to poor structure of the soil.

Depth of tillage is not an important matter in development of good tilth. The soil needs only shallow tillage to create a good seedbed. Plowing to 3 or 4 inches deep is sufficient for this purpose. Deeper tillage to 6 inches may be needed to incorporate residues. Tillage deeper than 6 inches is wasteful of energy, as the force needed to pull a plow through soil increases as the depth of tillage increases. Incorporation of fertilizers, limestone, manures, composts, or other soil amendments deeply in the soil may place these materials out of reach of plant roots, particularly during the early season when seedlings are becoming established. Deep tillage, 15 inches or so, to break up pans has only limited success. Hardpans may be fractured successfully by deep tillage with chisel plows. Clay pans may be separated by tillage but usually fuse together again.

A number of implements for tillage and practices for management of land by tillage are available as options to growers depending on the kinds of soils that are farmed and on the personal preferences of the growers.

IMPLEMENTS FOR TILLAGE

Moldboard Plow

The basic implement of tillage is the moldboard plow (Figure 9.1). This tool is efficient in turning of soil so that desirable soil structure results and residues are turned under. Its actions are somewhat similar to turning the soil with a hand spade. The surfaces of fields tilled by the moldboard plow are clean, so clean in fact that some organic farmers and others object to the lack of residues on the surface. The clean surface may expose the soil to erosion by wind and water. Another problem is that tillage to the same depth year after year by the moldboard plow can lead to compaction of the soil and formation of a plow sole by traffic of the implements of tillage. The moldboard plow is desirable for use in soils with high gravel contents to ensure good mixing. It is the implement of choice for tillage of clayey soils. The moldboard turns the clayey soil so that the shearing, lifting, and twisting actions help to accentuate granulation. Sod or other residues are turned into the furrow, and the shattered soil is exposed for further fitting by harrowing, disking, or dragging. If soils are tilled too wet, the pressure imparted to the soil by turning on the moldboard often leads to poor soil structure—cloddy conditions with clayey soils. With sandy soils, the moldboard plow may not be needed, and disking may be sufficient or even better than plowing.

Disks

Further fitting of the land for planting after plowing with the moldboard is accomplished by harrowing or disking to give good soil tilth. Disking should be moderate. Excessive stirring of the ground dries it and promotes oxidation and depletion of organic matter. Good soil structure can be destroyed by slicing and compaction. Disk harrows may be used also for the initial fitting of land, but if deeper tillage is desired a disk plow should be used.

Disk Plow

The disk plow (Figure 9.2) is employed commonly in tillage of land that is rough with stones, with outcrops of rocks, or with tree stumps. It may be used also in conventional tillage of land that is not rough. The disk plow leaves a lot of residue sticking through the plowed zone to the surface. This condition is preferred by some growers because of the effects that the residues have on prevention of erosion. However, the unincorporated residues may not be killed by disk plowing and may resume growth

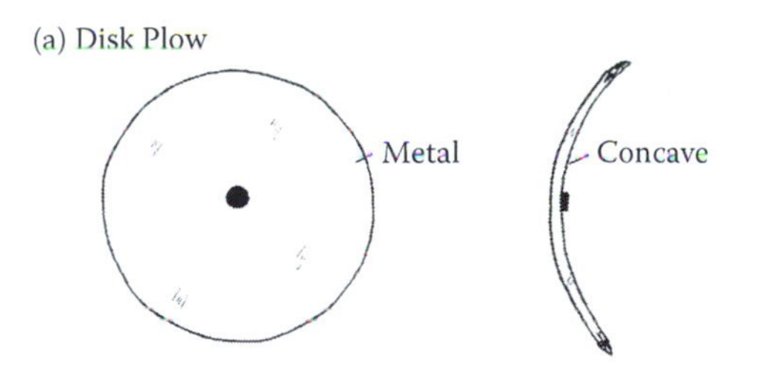

FIGURE 9.2 Diagrams of (a) front and side views of a disk plow and (b) appearance of land tilled with a disk plow.

as weeds. Further tillage operations by harrowing with disks or other implements often are needed to give control of these weeds.

Rotary Tillers

Rotary tillers are implements with rotating tines that cut into the soil. Depending on the nature of the soil and residues in the soil, a rotary tiller may prepare the soil in one pass across the soil. The rotary tiller has the action of a plow and a disk. Rotary tillers may present problems in use on wet, clayey soils. The tines of the tiller cut into the wet, sticky clay and throw slices that are clod-like. Drying of these slices without their degradation into a granular structure will produce poor tilth. Cultivation with rotary tillers in sandy soils should give few problems. Rotary tilling of land mixes organic matter in the tilled zone.

MINIMUM TILLAGE

Conventional tillage with the moldboard plow or other implements often leads to compaction of soil, increased soil erosion, and declines in soil organic matter contents. Research has shown that not only is it not necessary to plow the ground deeply but that the land does not have to be fitted as smoothly as once believed. Also, the entire surface of the land does not have to be prepared for planting. Practices that lessen the extent of tillage of land are called *minimum tillage*, *conservation tillage*, or *no-tillage* operations.

One practice of minimum tillage is to plow the entire plot of land conventionally and then to prepare only narrow strips for planting. The plowed area between rows is left rough, leaving this area favorable for infiltration of water, favorable for control of erosion, and unfavorable for growth of weeds (the rough seedbed is poor for weed germination). Fewer trips across the field or garden under this system of seedbed preparation result in considerable savings in energy, time, and money. Also, rows of crops may be planted in wheel tracks following the implements, which crush the clods and leave a seedbed suitable for planting without further preparation.

Another system of seedbed preparation involves plowing of only strips in which the crop is planted, leaving the areas between the rows untilled, sometimes with vegetation continuing in place (Figure 9.3). A zone is prepared with implements such as a rotary tiller or with plows or cultivators that are spaced to till strips and to leave untilled, inter-row zones. The zones between the row may be mulched conventionally, or a sod of grass, clover, or other plants may be grown as a *living mulch*. With employment of living mulches, the tilled zone for the row crop should be a minimum of 18 inches wide. Generally, some control such as mowing is applied to the inter-row plants (living mulch) to ensure that they do not become weeds and restrict the growth of the row crops. The mulched or mowed zone can make the field or garden attractive and provide a path for access to maintenance and harvest of the crops.

In no-tillage agriculture, only a zone for placement of the seeds or seedlings for transplanting is prepared. This zone need not exceed the width that is necessary to accommodate placement of the seeds or seedlings or the minimum width that

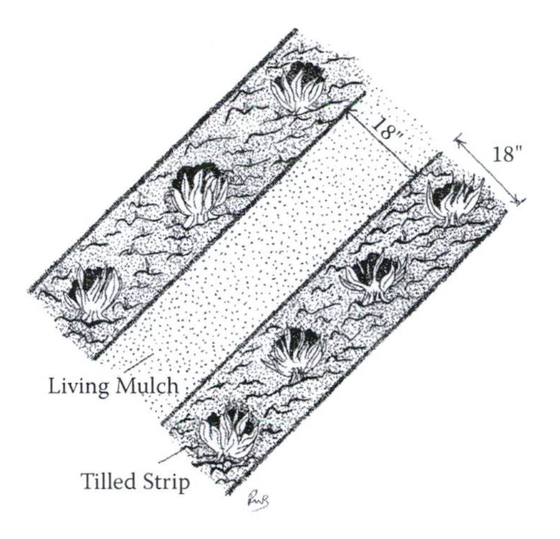

FIGURE 9.3 Strip tillage with 18-inch-wide cultivated zones for planting and a living mulch between cultivated zones.

can be tilled with implements (hoes, spaces, chisels, cultivators). Special multipurpose implements may be needed for large-scale operations. Use of these implements increases the amount of work that must be done in one operation; that is, tillage and planting are done in one operation. This requirement makes the practice of no-till planting unpopular with some growers. Also, no-till implements can be expensive, and farmers may not want to invest in this equipment. Till-planters (which perform tillage and planting in one operation) for large-scale farming can be rented from agencies that promote practices of soil conservation (conservation districts, the Natural Resource Conservation Service, and Cooperative Extension).

In large-scale, conventional, no-till operations, herbicides are used to kill weeds or cover crops, and the main crop is planted through the stubble and residue. The organic grower cannot use most herbicides in this manner but can develop systems in which no-tillage practices can be used. The organic grower can lay down an organic mulch for weed control on untilled land and then plant row crops through the mulch. The organic grower can also grow a tender cover crop, such as winter oats, which is fall seeded and which grows well in the fall before it is winter killed. The residue of the winter-killed crop remains as a cover of the ground in the spring. Crops may be planted through this residue without tillage other than in the strips in which the crops are planted. Crops that leave these winter-killed residues are referred to sometimes as *smother crops*, although this term usually refers to a dense-growing crops that stops the growth of weeds. Dense-growing crops, such as rye, clover, and vetch, that suppress the growth of other plants, particularly weeds, are smother crops. These crops can be used in the living mulch zones of minimum-tillage operations.

No-tillage systems are excellent for limiting soil erosion, for increasing organic matter and water in the soil, and for lessening soil compaction. The limited stirring of the ground lessens oxidation of organic matter and drying of the soil. Water is conserved by lessening of runoff and evaporation and by increasing infiltration.

Fewer trips with implements across the field lessen the opportunities for compression of the soil. However, under no-tillage management, concerns arise that surface applications of fertilizers and lime do not provide adequate nutrition of crops, and amounts of application are increased to ensure that nutrition is adequate. Occasionally, tillage is suggested to mix fertilizers and lime into the soil. However, rooting of crops in untilled soils is shallower than in tilled soils, due to more moisture near the surface, and the shallow roots may have access to phosphorus and potassium that move into shallow subsurface zones of the soil. Nitrogen fertilizers will move readily into the soil but also are subjected to losses if they remain on the surface of the soil. With time, lime will penetrate into the untilled soil, perhaps having an effect on soil pH to depths of 4 inches after 1 year. Amount of lime applied generally does not need to be increased for no-till systems above that required for tilled land.

Problems introduced by no-tillage farming include difficulties in control of plant pests and hindered warming of soil in the spring. Control of diseases and insects may present greater problems in untilled systems than in tilled systems because the residues on the soil surface give sites in which pests may be harbored. Warming of heavy (silty, clayey) soils is more likely to be slowed by no-till practices. In the spring, soils are colder under the surface-contained residues by as much as 6°F than in soils of tilled land. Cool temperatures may inhibit germination of seeds and establishment of seedlings. Cold temperatures may hinder the ability of plants to absorb phosphorus, leading to stunted growth because of the inability of plants to get enough phosphorus from cold soils. These problems do not occur as commonly in sandy soils as in silty and clayey soils.

Conservation Tillage

Minimum tillage is a form of conservation tillage, but conventional tillage can also be performed in ways that lessen soil erosion. *Contour farming* runs the rows of crops across sloping land rather than up and down the slopes. The contoured rows intercept and reduce the velocity of water moving down the hillsides, thereby lessening the movement of soil down the slope relative to movement that occurs with rows running up and down the hills. On cultivated steep slopes, strips of other crops or untilled land with cover may be alternated with row crops to further decrease erosion. The untilled strips may be permanent grass or trees or may be crops that are harvested for forage or pastured. On land that slopes more than 4% to 6% (4- to 6-ft rise or fall in 100-ft lateral distance), some form of conservation tillage should be practiced.

INTERTILLAGE OF CROPS

Intertillage refers to the cultivation between rows or within rows of crops. The most important purpose of this tillage is to control weeds. Intertillage is sometimes practiced to create a soil mulch for conserving water. Aeration of soil may be facilitated by tillage of soils that crust.

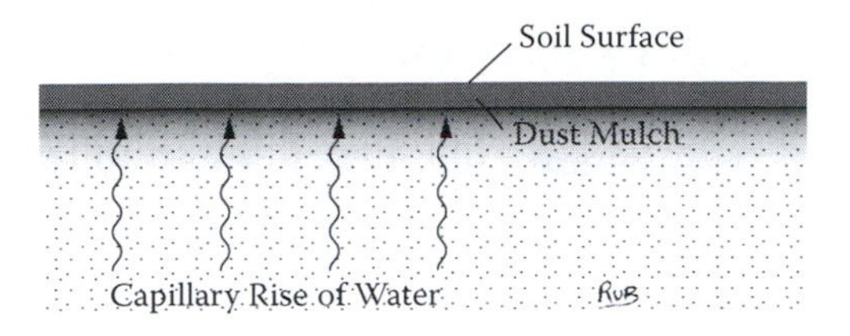

FIGURE 9.4 A dust mulch, which interrupts capillary rise and evaporation of water from a soil surface.

Weed Control

Shallow cultivation is all that is needed for weed control. Scraping of the ground is sufficient to cut off the weeds. Deeper cultivation dries the soil unnecessarily and may prune roots and reduce the capacity of the crops to obtain water. Routine cultivation on a fixed schedule should be avoided. Intertillage for weed control should occur when the weeds break though the soil surface. Tillage is usually necessary after a rain, which stimulates germination and emergence of weeds.

Soil Mulches

The benefit of a soil or dust mulch (Figure 9.4) to stop evaporation of water from the surface of soil is debatable. A soil mulch may have more benefits in dry areas than in humid areas. In humid areas, the mulch has to be recreated after each rain. A problem in dry areas might be enhanced wind erosion of the dry, powdery mulch. Stirring of the soil soon after a rain probably does not help to conserve water. In fact, the surface cultivation may accelerate drying though the enhanced aeration of the surface soil. Deep cultivation to 2 inches or more is not a good practice because of the hastened evaporation of water from the tilled zone.

Breaking of Crusts

Fine-textured soils, especially those with high silt percentages, may crust. Aeration of the soil may be impeded by the crust. Crusts can hinder emergence of seedlings of crops (Figure 9.5). Tillage to break the crust so that seedlings will emerge is a good practice, if not a mandatory one in some cases. Tillage to break crusts for the purpose of improving aeration may not be needed, and the severity of the problem must be assessed by visual inspection of the area before initiating tillage for this purpose. If seedlings cannot emerge through the crust, breaking of the crust likely is needed. Special implements, such as rotary hoes, are used to break crusts. Sands

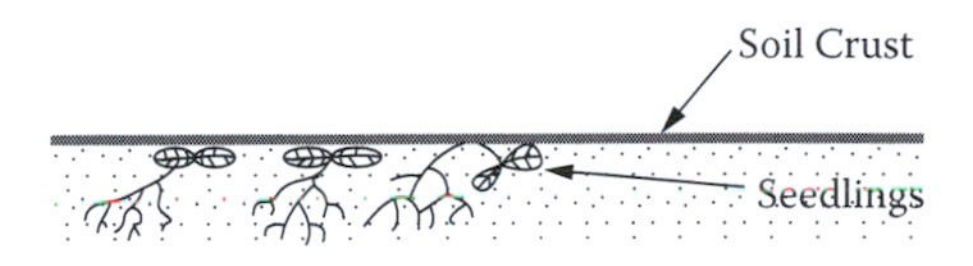

FIGURE 9.5 Inhibition of seedling emergence through a soil crust.

and loams generally do not present problems with crusting and need not be tilled for this purpose. Crusting of clayey soils is not of the concern as with silt soils because the clays crack and check and break up the crusts, whereas silts do not have this action.

10 Weed Control

The general definition of a weed is *a plant out of place*. More specifically, a weed is defined as a plant that is a nuisance, that is a hazard, or that causes injuries to humans, livestock, poultry, or crops. This chapter will concentrate on the control of weeds as pests in crop production. Weeds originate from wild species that are native to cultivated areas, from wild species that were introduced intentionally or by accident, or from species that are crop plants. The kinds of weeds that one deals with vary with the practices that are used in the production of crops.

Cultivation (tillage) of soil to grow crops disturbs the habitat in which wild plant species are growing. Tillage can destroy the weeds that are obvious on untilled fields. However, tillage can create an environment that is favorable for wild species to emerge as weeds in a crop. Sometimes, disturbing the ground allows dormant seeds to germinate, and plants will emerge that otherwise would not be obviously present. Also, cultivation of land may lead to further dissemination of weeds with the equipment or in other operations involved with crop production. Applications of farm manures and composts and grazing of livestock can spread weed seeds, as viable seeds are carried by each of these materials or vectors. Seeds of weeds growing in crops may be spread by harvesting or harvested with the crops and sown with the crop at planting in the next season. Seeds of crops left in the field after harvest may emerge as *volunteer* plants, which are weeds, in the following crop.

CROP DAMAGE FROM WEEDS

Weeds cause more losses in agricultural production than any other class of pests. Losses in developed countries in the temperate zones, in which weed control measures are efficient, range from 10% to 15% of the total value of field-grown agricultural products. Direct damage to crops by weeds occurs from the competition of weeds with crops for light, nutrients, water, and space. Faced with this competition, crops are unable to reach their full capacity to grow, and yields will be restricted. Some weeds inhibit crops through *allelopathy*, which is the effect that one plant has on another. Some weeds exude materials from their roots, or upon decomposition in the soil, weed residues release materials that inhibit crop growth by means other than competition. Some crops are allelopathic on weeds, and the roots or residues of these crops restrict growth of weeds or growth of other crops in rotations. Interference of weeds with crop production is a major cost in time and money, because the grower must spend a lot of time and devote a lot of resources to control of weeds.

Indirect loss of yields occurs through the effects of weeds on crop quality. Weed seeds mixed with crop seeds lower the value of the product for seed, feed, or food. Removal of weed seeds after harvest increases the costs of handling the crop, and as a result, the grower receives a lower price for weed-infested produce. Unripe seeds can increase the moisture content of crops, again resulting in a docking in price and in a lesser capacity of the crop to be stored without spoilage. Coarse-stemmed weeds in hay damage its quality as a feed and deter its consumption by livestock. Livestock grazing on weeds may be sickened or killed by poisonous plants. Weeds can give off-flavors to milk produced by livestock that graze on weedy pastureland. Weeds can block grazing of forage plants that are overgrown by canopies of weeds. Seeds of weeds may be attached to the fur, tails, manes, or ears of livestock and present problems to the livestock and to the grower. Weeds make harvesting of crops difficult, for the stand of weeds may be an impediment to the movement of equipment or people through the field, may clog the harvesters, or may damage the equipment.

Weeds are also alternate hosts for crop diseases, provide places where insects and diseases may harbor, and provide means of perpetuating diseases and insects in land. Weeds cause enormous problems in waterways, lakes, and ponds. Weeds clog the movement of water in irrigation ditches. Weeds can restrict flow or water in streams and lead to silting.

Health of humans, livestock, and pets is affected by weeds that are poisonous plants. Animals may be injured by ingestion of poisonous plants or may suffer rashes or other skin disorders from contact with the plants. Pollen from weeds is a common cause of allergies.

CLASSIFICATION OF WEEDS

Weeds can be classified broadly as *annuals*, *biennials*, and *perennials*. Annual weeds complete their life cycles in 1 year; that is, they grow vegetatively, produce flowers and seeds, and then die usually in one growing season. Strict *summer annuals* have all of their growth activities in one crop-growing season and have no means of survival from one year to another except by production of seeds. *Winter annuals* germinate in the fall, pass through the winter as seedlings, and produce most of their growth and seeds in the next growing season. Winter annuals often may be perceived as being *biennials*.

Annual weeds are the most frequently encountered weeds. Common summer annual weeds are crabgrass, foxtail, lambsquarters, pigweed (amaranthus), common chickweed, ragweed, purslane, morning glory, and galinsoga. Some winter annuals are common chickweed, wild mustard, shepherd's purse, and annual bluegrass. Some of these species behave as summer annuals as well as winter annuals.

Biennial weeds complete their life cycles in 2 years. In the first year, the growth is only vegetative as rosettes, and in the second year, the plants grow vegetatively and then flower and produce seeds. Wild carrot (Queen Anne's lace) and wild mullein are common biennials.

Perennial weeds live vegetatively in the soil for 2 or more years. They may flower and produce seeds in the first year. Perennial weeds often have some fleshy storage organs (roots, tubers, rhizomes, bulbs) that enable them to live from one year to another and often to propagate asexually. Perennial weeds because of their storage organs are harder to control by cultivation than annual or biennial weeds. Some common perennials are dandelions, docks, quackgrass, thistle, milkweed, nutsedge, and johnsongrass.

CONTROL OF WEEDS

In crop production, some method of weed control usually is needed to protect the crop. The purpose of weed control is to impede growth or to reduce the stand of weeds so that they do not imperil crop growth. A general aim is toward control of 95% of the weeds in the crops. Higher levels of control are desirable or possible but often are not practical. Lower levels of control do not offer sufficient protection to crops. Weed control in the early part of the life of a crop is important. Young crops may not be competitive with weeds and will be overwhelmed with weed growth. Control of weeds is essential during the first 6 to 8 weeks of crop growth. During this time, the crop can produce enough top growth to give a canopy that shades and inhibits weed growth. Weeds under the canopy may not be competitive with the crops. With some crops, these weeds can continue to grow, even though in sparse populations, and finally give difficulty in harvesting of crops.

General Methods of Weed Control

Mechanical

Mechanical control includes hand weeding, hoeing, cultivation, mowing, clipping, and burning. These practices remove weeds or damage them sufficiently so that the crop has a competitive advantage.

Cropping

Rotations of crops change environments in which weeds grow. Weeds often become established with a certain crop, and rotation of crops changes the scene in which weeds compete. Densely seeded crops or those with dense canopies have strong capacities to compete with weeds. Fertilization often helps crops to out compete weeds, and the competitive response of crops to fertilizers will vary, hence giving variable measures of control with crop rotations. Rotations of crops as a weed-control practice are addressed specifically later in this chapter.

Biological

Biological control involves introduction of an organism to destroy they weeds. Insects are the most common organisms used in this respect. Concerns always exist as to whether or not the insects introduced to eat weeds will turn to eating the crops after the weeds are eliminated. Weed-eating insects are not commonly available for growers to utilize in crop production. Reclamations of range land from weed infestations in Australia and in western United States (see latter sections of this chapter) are often-cited examples of successful use of biological control. Farm animals also may be used in biological control. Weeds may be grazed by livestock after crops are harvested. Sometimes livestock have been used in weed control in tree plantations. Geese may control weeds in strawberries.

Chemical

Selective herbicides are used to control weeds in crops. The selection of herbicides available on the market is very large. However, the organic gardener and farmer have few, if any, options for herbicides that are allowed in certified organic practices. Some herbicidal soaps and organic acids are marketed and are certifiable as organic, weed-control practices.

Specific Methods of Weed Control

Cultivation (Tillage)

Conventional tillage with the moldboard plow controls weeds by cutting them in the soil and burying the shoots in the overturned soil (*furrow slice*). Disk plows and harrows give less control than the moldboard plow because of incomplete burying of the shoots. With minimum tillage or no-tillage systems, herbicides or mulches often are used in connection with preparation of the land for planting.

Intertillage is a primary means for organic control of weeds in a growing crop. Weeds are removed mechanically between the rows by cutting at the soil surface (scraping) or by shallow tillage in the soil. Cultivation needs to be shallow not to damage roots of crops. For most crops, scraping of the soil is as beneficial as shallow cultivation to 1 or 2 inches deep. Few crops benefit from the dust mulch that is created by shallow cultivation. In fact, in humid regions, the tilled zone will dry, and water available to plants will be lessened by evaporation of water from the tilled zone. Intertillage cultivation should be as soon after weed emergence as possible rather than following a fixed schedule. Weeds within the rows of crops are difficult to control. These weeds may need to be removed by hand or by hoeing. Mechanical cultivators with disk coulters (listers) may be set to bury these weeds by throwing soil over them during cultivation. Rotary hoes also may remove weeds from between the rows when the crops are young but better anchored than newly emerged weeds.

Conventional tillage and intertillage can reduce potential weed populations by 10% to 15% per season by killing newly sprouted seeds, thereby reducing the number of weed seeds in the soil, provided few or no weeds are allowed to form seeds during the same growing season. Occasionally, annual weed populations are so high that conventional tillage cannot control weeds sufficiently for crop production. Also, perennial weeds may infest a field so severely that crop production is not possible with conventional tillage. In these cases, *fallow tillage* must be practiced. In control of annual weeds, fallow tillage involves cultivation of uncropped (fallow) land, usually by disking or comparable means, to control virtually all annual weeds as soon as they emerge. In control of perennials, fallow tillage allows for exhaustion of the stored reserves of these weeds. In this case, tillage occurs as often as new growth emerges from the storage organs and before enough growth occurs to replenish the reserves in the organs. Fallow tillage also allows for desiccation of stolons by exposing them to the atmosphere. With perennials, tillage typically is on a 1-week or 2-week cycle.

Fallow tillage may take land out of production for a growing season or for most of the season depending on the severity of weed infestation. In some cases with perennial weeds, fallow tillage may be required for 2 years or more to achieve sufficient control. Weed seed populations of annuals in soil typically are reduced by 30% to 50% per year by fallow tillage compared to the 10% to 15% reduction by conventional tillage with crops present. Control of perennial weeds may appear better than that achieved for annuals, since the objective in control of perennials is to destroy the infestation of adult plants rather than to reduce the weed seed population in the soil. Care must be taken to avoid letting perennials go to seed and, during cultivation, to avoid spreading of plant segments from which new plants can arise.

Burning

Burning of gardens, fields, or pastures usually has little value in long-term weed control. Burning gives a temporary set-back by clearing away debris and allowing other plants a chance to compete. Burning of pastures is used to clear dead grass that may impede livestock from grazing on tender young forage that is emerging. Burning may be practiced to clean a field that is covered with dense stands of brambles or dead grasses, which would interfere with tillage by clogging the implements or by failing to be buried adequately by the tillage operation. Burning usually does not kill storage organs or crowns to stop regrowth after the dead shoots are burned away. Seeds in the soil usually can survive burning over the surface. Temperatures of 175 to 212°F for at least 15 minutes are necessary to kill seeds in the soil and near the soil surface and to kill crowns. Unless a deep mat of organic debris covers the soil, these temperatures will not be achieved. Burning generally requires a permit from local governments and will be restricted to limited periods during a year.

Flaming with torches is a burning practice for weed control. Propane-fired, hand-held torches can be used to kill weeds on small plots of land, in and along sidewalks and driveways, and other easily accessible small areas. Flaming should be used on green vegetation to avoid fires. Implement-mounted torches are used to burn weeds within rows of crops if the weeds are tender and more susceptible to injury than the crop plants. Generally, flaming burns away or wilts the topgrowth of weeds. Annual weeds will be killed. Perennial weeds are likely to be return from the same stock after burning. Seeds in soil will not be damaged by burning of topgrowth, and flaming of the soil has little effect on the weed seeds in the soil.

Mowing or Clipping

Mowing (cutting plants close to the ground) or clipping (removal only of the upper portion of the plants, for example, flowering portions) helps to increase the competitive nature of crops. These operations control vegetative growth of weeds and keep weeds from going to seed. Weeds in lawns and other turf areas, pastures, vacant lots, and open spaces often are controlled by mowing or clipping. Mowing or clipping greatly improves the appearance of open spaces.

Crop Rotation

Certain weeds often are associated with certain crops—for example, cocklebur, lambsquarters, morning glory, crabgrass, and other summer annuals grow with cultivated row crops, such as corn, soybeans, and vegetables; onions, garlic, and mustards grow with wheat; burdock, broomsedge, ragweed, thistles, and mullein are common in pastures. Changing of crops will help in control of weeds in these associations. Sometimes, however, crop rotation permits growth of weeds that otherwise would not appear, such as those that emerge when pastures are disturbed by tillage and put into production of row crops. In these cases, another practice of weed control must be used during production of the first crop or two on the land. Then, rotation can be effective for weed control in subsequent crops.

Mulching

Mulching (see Chapter 8) helps in control of weeds by providing a soil cover that is too thick or too impervious for weeds to penetrate. Mulching is a particularly good practice when labor is not available for continuous control of weeds by other means throughout the season. Straw is a good mulch for weed control. The layer of straw should be at least 2 inches thick. Hay will work also. Hay with mature seeds, however, may increase weed seed population in the soil and give problems in future seasons. One bale of straw or hay will cover about 50 square feet of area. A few sheets of newspaper underneath the straw or hay increase the level of weed control achieved by these mulches. A 2-inch or thicker layer of sawdust, woodchips, or bark gives good control of most annual weeds. Grasses, chickweed, and perennial weeds may not be controlled fully by mulches, but their control by mechanical means will be made easier. Sawdust, bark, or woodchips should be used in perennial plantings and not in areas that are to be tilled at the end of the season. Incorporation of woody materials into the soil will lead to immobilization and subsequently to deficiencies of plant nutrients, especially nitrogen. The woody mulch on top of the ground does not present problems with immobilization of nutrients, as the mulch is not in as extensive contact with the soil as incorporated sawdust, bark, or woodchips would be. Mulching with manures or composts gives excellent control of weeds, especially if a layer of newspapers is placed under the mulch. However, caution must be taken with the use of these mulches. Some manures and composts are ridden with weeds seeds, and weeds may grow from these seeds directly in the mulch or in the soil. The weed seeds in manures arise from the bedding for the livestock, from seeds of hay that may enter the livestock waste and that gets mixed with the manure, or from seeds of the hay or weeds in the hay that pass through the livestock without digestion. Composts may contain weed seeds from the materials that were composted. Often weeds grow and go to seed in uncovered compost piles. In some cases, the nutrition added by the mulch of composts or manures stimulates growth of weeds from seeds borne in the soil so that weeds are more of a problem than in unmulched soil. The layer of newspapers underneath the mulch usually inhibits the emergence of soil-borne weeds through the mulch. Straw or weed-free hay make good covers over the newspaper. When used in combination with the newspaper mulch, the compost, hay, or straw need only be thick enough to cover the paper for good appearance or to hold the paper in place. Without the newspaper mulch, these materials should be applied at least 1.5 inches to several inches thick.

Black plastic mulch gives good weed control as well as warming of the soil. Pulp from shredded paper or vegetable garbage dries to an impervious layer that controls weeds. The pulp can be made by running the paper or vegetation through a garden shredder. Mulches of rocks, sand, and gravel control weeds in uncultivated areas but not without some problems. Weeds that grow around or through the rock mulch or through the sand or gravel may be difficult to control without disruption of the mulches. Handweeding, burning, or herbicides may need to be applied to control weeds in rock, sand, or gravel mulches. Dust mulches created by shallow tillage give some control of weeds until rainfall destroys the mulches.

Fertilization

Fertilization may increase the competitive capacity of a crop and enable it to outgrow weeds. The shading from the canopy of a vigorous crop will slow the growth of weeds. A well-fertilized crop, especially if given some aid by mechanical weed control measures during its early period of growth, will have a good competitive advantage over weeds after the crop canopy is established.

Sometimes fertilization gives weeds a boost so that they outcompete a crop. With perennial crops, this effect might be noticeable particularly in the first year of fertilization, but in subsequent years, the crop will have the advantage.

Biological Control

Little interest has been generated for wide use of biological control of weeds because of the fear of hazards arising from the organisms used in the control. Most successes in biological control of weeds have been achieved with insects. Use of insects in biological control of weeds has had limited application in commercial agriculture in tilled land, and most successes have been achieved on large expanses of grazing land in relatively undisturbed conditions. Very successful applications with insects occurred in control of prickly pear cactus in Australia (ca. 1925) and control of Klamath weed in California (ca. 1944). In each of these cases, large acreages of pasture land were infested to the extent that cattle could not graze because of the dense stand of the weeds. The prickly pear was an added hazard because the cattle often would get cactus thorns in their lips and tongues and would starve from their inabilities to graze with these injuries. The Klamath weed gave the added hazard of being poisonous to livestock. In both of these cases, use of herbicides was not considered feasible because of the large expenses involved and the relatively low value of the land. Each of these weeds was virtually eliminated within 10 years by weed-eating insects that were specific for the prickly pear or Klamath weed.

Although no system is commercially developed yet, research is underway presently to evaluate biological control of weeds with insects, mites, nematodes, snails, fungal diseases, and other plants in agricultural environments on cultivated land and forests.

Livestock, poultry, and wildlife have been used in biological control of weeds. Geese will eat weeds in cotton, strawberry, and mint and not consume the crops. Cattle have been used to control weeds in Christmas tree plantations. Goats are reported to eat poison ivy. Cattle and hogs often are allowed into fields of harvested crops to forage on vegetative residues and grain and to graze on the wild grasses and other plants that emerge after harvest of the crops. Wildlife will help to control weeds in pastures and rangeland, although in many cases wildlife may be pests because of their competition with livestock and because of the damage that wildlife do to crops.

Allelopathy

Allelopathy refers to the effects that one plant has on the growth of another plant and can be employed in biological control. Black walnut trees control vegetation by a chemical that is derived from the leaves of the trees or from the roots. People should avoid planting of shrubs or small fruits underneath walnut trees. Many agricultural crops produce allelopathic chemicals that affect the growth of other plants. Sunflower seeds are allelopathic, and ground around bird feeders may be barren as the result of the effects of sunflower seed or hulls on growth of other plants. Some common agricultural crops are allelopathic while they are growing or leave allelopathic residues. Living sunflower plants inhibit growth of weeds. Decaying residues of wheat, rye, corn, sunflower, and soybean have allelopathic effects that help to control weeds. Residues of these crops may be incorporated from a previous crop for allelopathic weed control. The residues also may be grown off-site and imported and applied as mulches or incorporated into the soil for the benefit of allelopathic control of weeds. Growers should know that these residues may interfere also with the growth of crops and must take this effect into consideration in planning crop rotations and in green manuring. Weeds also have allelopathic effects and may affect establishment of crops. Foxtail, lambsquarters, and barnyardgrass are allelopathic weeds, and failure to control these weeds in one season may hinder the growth of a crop in the following season.

Herbicides

Few herbicides are available for organic production of crops. Perhaps aqueous extracts could be made from allelopathic crops and applied in attempts to control weeds, or the residues of the allelopathic crops could be applied for their herbicidal action. These procedures are relatively untested or expensive especially for large-scale operations. In the past, Stoddard's solvent (carrot oil) was considered to be an organic herbicide. This material was a dry-cleaning fluid also used for weed control in carrots and parsnips. It is not readily available in the marketplace. Calcium cyanamide and ammonium sulfamate are mentioned in some texts as organic herbicides. These materials decompose within 4 to 6 weeks to urea (from calcium cyanamide) and to ammonium sulfate (from ammonium sulfamate), which are nitrogen fertilizers. Since neither urea nor ammonium sulfate are organic fertilizers, it is not reasonable to consider calcium cyanamide or ammonium sulfamate to be organic herbicides.

Herbicides have been derived from organic acids and other constituents of plants. The controlling action of these materials against weeds is dehydration of foliage. Young, soft weeds are more susceptible to control than mature weeds. Herbicides derived from organic acids are not selective and can damage crops as well as kill weeds. Uses include spot-killing of weeds and cleanup of weeds in sidewalks and driveways and in rock, gravel, and sand mulches. Acetic acid (vinegar), citric acid, pelargonic acid (geranium), limonene (oranges), and clove oil or cinnamon oil (eugenol) are commonly available herbicides, some of which have organic certification. Some common household soaps and detergents are phytotoxic and might be considered for weed control.

11 Insect Control

Insects are numerous invertebrate animals of the class Insecta. In the adult stage, insects are characterized by having six legs (three pairs) and a segmented body that is usually winged. Various similar invertebrate animals, such as spiders, mites, ticks, millipedes, and centipedes, which are not taxonomically insects, are defined loosely as insects in common terminology by the public. Control of these organisms often is considered under topics of insect control. Pathogenic nematodes, which are microscopic worms, usually are studied as plant diseases.

The developmental life cycles of insects commonly undergo changes in form called *metamorphosis*. The cycle for all insects begins with an *egg*. The egg hatches into an immature insect stage called a *larva*. The larvae are known commonly as worms, grubs, caterpillars, maggots, or nymphs depending on whether the final stage is called a *moth*, *wasp*, *butterfly*, *fly*, *bug*, *beetle*, or perhaps other name. As the larvae grow, they feed heavily on their hosts. Larval stages of insects often are the most ruinous to crops because of the heavy feeding to support the growth of these organisms. Larvae have limited capabilities to travel and spend their life stage on one or a few plants. Larvae shed their skin several times until they reach maximum size. At the end of the growth of the larval stage, many insects form a legless, compact stage called a *pupa*. The pupal stage is a sort of resting or dormant stage and has a hard protective outer case, sometimes with a cocoon-like structure. The *adult* grows from the pupa, and when mature, the adult splits from the pupal case. The adult is a fly, a moth, a butterfly, a beetle, or other winged or unwinged form of the organism. Winged adults have considerable capacities to travel. The feeding habits of the adults may resemble or differ widely from those of the larvae. Control of insects requires that the grower recognize whether the larvae or adults are the main feeding pests of plants.

Insects that undergo the changes in form of *egg-larva-pupa-adult* have a *complete metamorphosis*. Some insects do not have a pupal stage. The larvae of these insects look like miniature adults, except wingless. These larvae are called *nymphs*. The nymphs grow and molt, and with each successive molting, the nymph stages more closely resemble the adult stage. With the final molt, the insects emerge as fully formed adults. Insects that undergo changes of *egg-nymph-adult* have an *incomplete metamorphosis*. Knowledge of the life cycles of insects is important in understanding the feeding habits, the migratory habits, and the damage by the insect to crops. All of these factors are important in development of procedures for insect control.

CROP DAMAGE BY INSECTS

Direct damage is done by plant-eating insects. These insects chew, suck, or bore into buds, leaves, stems, roots, flowers, fruits, or seeds. Damage to the vegetative portions occurs from defoliation, weakening of the stems, or destruction of the roots. Plants may be killed by this damage. As a rule, the earlier the damage occurs to the plant, the more ruinous the damage is. Damage to flowers, fruits, and seeds occurs from boring insects in the larval stage but also from adults that may feed on buds or immature fruits. Products of these damaged flowers, fruits, or seeds usually are unmarketable because of their malformed shapes or their containing the living larvae or adults. Often, eggs or larvae persist into the harvest and stored product and emerge into larvae or adults in the crib or bin or even in the kitchen cabinet.

Indirect damage by insects is a major concern in crop production. Insect damage weakens the plant to further damage by wind, rain, or plant diseases. Diseases often invade the infested area to feed from the feces of the insects or on the damaged regions of the plant. Aphids feeding on leaves cause a leaking of a sugary substance, honey dew, that is feed for black molds that ruin the appearance of plants. Insects, such as aphids, leafhoppers, and cucumber beetles, are vectors of plant diseases. One means of spread of viral diseases, yellows diseases, and wilts or blights is by insects. Dutch elm disease is spread by the bark beetle.

METHODS OF INSECT CONTROL

Only organic means of insect control will be presented in this chapter. These methods include use of *sprays* and *dusts,* introduction of *biological control* measures, and application of cultural practices. Insecticidal sprays and dusts are materials that kill insects or limit the activity of insects through chemical toxicity or physical action on the organisms. Sprays and dusts may differ little from one another except in mode of application with one being applied in a liquid form and the other in a dry form. Biological control involves the introduction of predators, parasites, or diseases that feed on or infest insects. Cultural practices include management procedures that growers can employ to limit the access of insects to crops.

Application of Organic Sprays and Dusts

These pesticides are chemicals derived from mineral deposits, from plants, from animals, or sometimes by manufacturing from natural ingredients. Whether organically derived, all sprays and dusts should be handled with care, as they may present hazards to health. Usually some characteristic, such as abrasive or toxic action, is the factor that allows sprays or dusts to be insecticidal. Organic sprays and dusts can be as injurious to humans and other animals as chemically manufactured pesticides. Often, the distinction between organic and nonorganic ("chemical, manufactured") pesticides is marginal or weakly defined. Both types of insecticides should be treated the same. *Directions provided with the materials should be followed, and records of application should be kept. The pesticides should be used only on the crops specified, in amount specified, and at time specified in the directions. The pesticides should be applied at the most susceptible stage of the insect.*

In applications of insecticides, plants should be sprayed or dusted thoroughly, being sure that the pests come directly in contact with the pesticide or its vapors. Pesticides should be applied in calm weather to lessen blowing and drifting. Application is recommended to be in the cool of the morning or evening. This practice helps to avoid injury to plants and to maintain the efficacy of the materials, protecting against their inactivation by heat, light, or drying.

The applicator should prepare only as much spray as is needed for treating of the crops. Excess material should be diluted with water and dumped away from all water sources. Pesticides, containers in which pesticides were held, and spraying and dusting equipment should be kept away from children, pets, livestock, poultry, and wildlife.

The applicator should avoid inhaling or swallowing the pesticide and should avoid contact of the materials with the skin. Proper equipment should be used to apply the pesticides. Protective clothing, goggles, or masks may be needed. Care should be taken to avoid consumption of the pesticides on products that have been treated with sprays or dusts. Pick produce before the pesticides are applied, wait the appropriate time before harvesting treated produce, and wash treated produce if possible before consumption. These rules apply to purchased or to homemade pesticides.

Many of the materials described here as sprays also can be applied as dusts if the initial material is dry. Sometimes the insecticide for spraying is sold in a more concentrated formulation than the dust. Growers should check with certifying agencies to determine the usage of these products in organic agriculture.

Sprays (See Table 11.1)

Water

Water alone can be an insecticide. Wetting of insects may cripple them or limit their movement on or onto plants. Wetting of hairy insects is facilitated by adding a few drops of soap solution or detergent. If water does not have an insecticidal action from these disabling injuries, it can help to lessen damage to plants by knocking the insects off plants. Insects can be washed off house plants and down a drain. Insects may be washed from bark or nests of insects by streams of water directed onto the plants. These effects of water are strongest against nonflying insects, such as aphids, cabbage worms, and red spiders.

Insecticidal Soaps

Soaps or detergents lessen the surface tension of water, allowing water to spread smoothly over the surface of leaves or insects and thereby improving the effectiveness of water as an insecticide. Soaps and detergents also are insecticidal. Dishwashing

TABLE 11.1
Sprays for Insect Control by Organic Means

Water	Alcohol	Sulfur
Insecticidal soaps	Diatomaceous earth	Milk
Ammonia	Salt solutions	Whitewash
Mineral oils	Starch	
Neem		

detergents are insecticidal but also are phytotoxic. With most certifying organizations, only soaps are considered to be organic pesticides.

Care must be taken not to use more than a few drops of soap or dishwashing detergent per gallon of water, that is, no more than 20 drops (about 1 ml) as a starting point, The phytotoxicity of these materials should be evaluated on a portion of a plant before widespread applications are made. Soaps or detergent solutions may collect in whorls or other areas of plants and concentrate upon evaporation to toxic levels. Plants with thin cuticles are more sensitive than ones with thick cuticles. For example, garden beans, cucumbers, and peas are sensitive to soaps and detergents, tomatoes and potatoes are less sensitive than those crops, and waxy plants such as cabbage are very tolerant of these sprays. Either soap or detergent solutions may be used as herbicides if the concentrations of the material are high enough in solution.

Insecticidal soaps are available on the market. Generally, these soaps are made from naturally occurring fatty acids. Diluted insecticidal soaps have low phytotoxicity and low mammalian toxicity and can be used indoors or outdoors. No residues are left to have long-term environmental effects. Soaps can be applied to crops up to harvest of produce. Soaps, or detergents, are effective against most soft-bodied insects, including aphids, spider mites, caterpillars, leafhoppers, whiteflies, mealybugs, and lace bugs. Soaps are less effective on scales and beetles than on soft-bodied insects. Beneficial as well as harmful insects can be killed by soaps or detergents. The effectiveness of the soaps or detergents comes from their action on the cuticle and cell membranes of insects. Cells in contact with the soaps or detergents become leaky, resulting in dehydration and death of the insects.

Detergents are manufactured materials and are not considered to be organic substances. In fact, addition of detergents to certain potting media as wetting agents prohibits the use of these media in organic agriculture. Hence, use of detergents as insecticides is not likely to receive organic certification.

Ammonia

Household ammonia, sudsy or not sudsy, is as an insecticide. Ammonia is toxic to insects and plants. The household ammonia should be diluted about 4 oz. in a gallon of water to be applied as a spray. This mixture should be tested on a few leaves of the affected plants before it is used on the whole plants.

Mineral Oils

Highly refined mineral oils are applied as emulsions in water and are an organically acceptable chemical control. Oils coat bodies of insects and suffocate them. No insect is known to have resistance to oils. Oils kill eggs and are very effective against scales, which otherwise are difficult to attack with insecticides. However, oils may injure foliage. *Dormant oil* should be used on leafless shrubs or trees or on some conifers, because it may defoliate broadleaf trees or shrubs. Dormant oil is known also as *Volck's oil*. Sometimes the oil is marketed mixed with sulfur. The sulfur increases the effectiveness of the oil as an insecticide and also gives fungicidal properties to the mixture. *Summer oil* or *white oil* is more refined and lighter weight than dormant oil. Summer oil can be used on leafed-out trees and shrubs but should be evaluated to ensure that phytotoxicity is not imparted by the spray.

Alcohol

Rubbing alcohol is marketed as a 70% aqueous solution of isopropyl alcohol. Scale insects can be wiped from plants with a cloth or cotton ball wetted with the undiluted alcohol. Insects such as mealy bugs can be daubed with undiluted alcohol with cotton applicators on a stick. Alcohol sprays can be prepared from 8 to 16 fluid ounces diluted into a quart of water, a dilution of one part alcohol to two to four parts water by volume. In either of the cases of using undiluted alcohol wipes or daubs or diluted alcohol sprays, precautions should be taken not to injure foliage. A small area or a few leaves should be tested with the alcohol before the whole plant is treated. Alcohol is used against pests such as aphids, scales, and whiteflies.

Diatomaceous Earth

Diatomaceous earth is silica derived from fossilized marine algae called diatoms. These algae are single-celled organisms that have an outer shell of silica. Diatomaceous earth has many industrial and home applications, with common uses being in polishes and swimming pool filters. As an insecticide, it is a nonselective, abrasive material with particular effectiveness against crawling, soft-bodied adult insects, caterpillars, snails, and slugs. The particles of silica are microscopic and needlelike. The particles penetrate and abrade the waxy coating (cuticle) of insects, causing them to dehydrate. Diatomaceous earth is considered nontoxic to mammals, but if it is inhaled, it can be very hazardous by irritating the mucous membranes. Dust masks should be worn while it is being handled. The material can be dusted around the base of plants to control root maggots and soil-dwelling insects. It can be dusted on leaves to control chewing insects. Dew or irrigation water on the foliage helps to retain the diatomaceous earth on the plants. It can be applied as a spray in water instead of as a dust. A few drops of soap or detergent will improve the wetting action of the spray.

Salt Solutions

Table salt (sodium chloride) dissolved in water can be an insecticide that is effective against soft-bodied pests, such as cabbage worms, aphids, and spider mites. The insecticidal action of salt sprays is through their desiccating of the insects. Dissolve about 1 oz. of salt in a gallon of water to prepare a spray. This amount of salt in solution should not injure plants, but young plants may need to be tested on a few leaves before the whole plant is sprayed.

Starch

Starch sprays work by gumming up the leaves and trapping insects in place or by gumming up the insects if they are sprayed directly. Baking flour or potato starch dextrin can be prepared with a few tablespoons (2 to 4) per gallon of water or can be applied as dusts.

Sulfur

Sulfur is a naturally occurring mineral. It has insecticidal and fungicidal properties. Finely ground sulfur is wettable and can be sprayed. Commercial products are available readily. Sulfur can be applied as a dust as well as sprayed. Sulfur might be applied in closed environments by vaporizing the sulfur from a heating element, but the safety and efficacy of this practice are questionable. Dusts may have clay or talc added to improve the dusting properties of the material. These specially prepared sulfur dusts would not be useful as wettable powders. Sulfur is a nonselective insecticide but is particularly effective against mites. Sulfur is also moderately toxic to mammals, fish, and other animals. Masks and protective clothing should be worn when applying sulfur to plants. Avoid use of sulfur in hot weather, because crop damage may occur. Sulfur is corrosive to equipment due to its oxidization and reaction with water and oxygen to form sulfuric acid.

Milk

Milk has insecticidal properties due to its proteins (globulins). Any kind of milk can be used. It is effective against soft-bodied insects such as cabbage worms. Milk also has some effects against fungi and viruses.

Whitewash

Whitewash is slaked lime (calcium hydroxide) prepared as a slurry in water. It is applied as paint on trunks of trees to protect against boring insects. Generally, whitewash should not be applied to foliage, unless it is prepared as a dilute spray that does not leave a heavy residue. Whitewash is fungicidal and offers protection against plant diseases when it is applied to plants. Storage rooms and root cellars often are whitewashed to control diseases that develop on stored produce.

Nicotine

Nicotine is an alkaloid from tobacco and is a powerful insecticide, one of the most toxic of the botanical insecticides. It is effective against most insects. It can be applied to soil or to foliage. A wetting agent (soap or detergent) in the spray helps with coverage of foliage. The fumes from nicotine are effective in control of insects, such as aphids, whiteflies, leafhoppers, and thrips that may be on the undersides of leaves or in the whorls of plants where liquid sprays may not reach. The volatility of nicotine allows it to dissipate quickly, although one recommendation with nicotine dusts is to use only on young plants to ensure no toxic residues at harvest. Nicotine sprays or dusts are or were sold as nicotine sulfate, which has a lesser mammalian toxicity than straight nicotine. Applicators should always check the label on the insecticide to ascertain permitted use of nicotine. Applicators should not remain in the area in which nicotine is applied especially in enclosed areas. Nicotine or nicotine sulfate, the insecticidal formulation, is highly toxic, and its use should be avoided by home gardeners. It should not be used on pets or farm animals. Nicotine that is distilled to prepare sprays should have no tobacco mosaic virus, but again users should consult the label to determine effects on solanaceous crops, such as tomatoes, peppers, eggplants, and potatoes. Dusts prepared from stems and leaves of tobacco and *nicotine teas* are likely to have the mosaic virus. Nicotine teas are prepared from aqueous extracts of tobacco products, for example, a cup of cigarettes or cigars in a gallon of water.

The availability of nicotine formulations is limited in the United States to imported products. It is no longer registered commercially for insect control. Nicotine sulfate is prohibited for use in organic agriculture by the U.S. Department of Agriculture (USDA) National Organic Program.

Neem

Neem or neem oil is extracted from seeds of the neem tree (*Azadirachta indica*), which is common in Africa and India. Neem is related to the chinaberry tree (*Melia azadarach*), which grows in the southern United States. Extracts of seeds of either species have insecticidal properties. The extracts have repelling, hormonal, and toxic effects against a broad spectrum of insects in larval or adult stages. Plants sprayed with neem extracts are unpalatable to insects. Neem is sold commercially. Neem has been reported to have fungicidal properties as well as insecticidal properties. Neem has a low toxicity to mammals. It is offered commercially as an alternative to nicotine sulfate.

Limonene

Limonene is refined from citrus oils that are extracted from peels of oranges and other citrus fruits. It is generally regarded as safe by the U.S. Food and Drug Administration and is used widely as flavorings and scents in foods, cosmetics, perfumes, soaps, and degreasers. Limonene has low oral and skin toxicities to mammals and dissipates rapidly from surfaces, leaving no residual effects. Limonene is a contact poison to insects but may have fumigant activity. Limonene is used more commonly against external pests of pets to control fleas, lice, ticks, and mites. Commercial products are available as sprays, aerosols, shampoos, aerosols, and dips. Limonene has herbicidal activity. It affects the cuticle of plants and imparts desiccation. Its use against insects on plants may be limited by this property. However, phytotoxicity of limonene is low.

Sprays Prepared from Plants

Many plants contain materials that are insecticidal and that can be extracted with water. Preparations are made by chopping leaves in a blender or by hand with soaking of the homogenate or chopped leaves overnight in water. Generally about 1 cup of fresh leaves is used per cup of water in which the leaves are blended or chopped and soaked. The blend is strained through cheesecloth or otherwise filtered to create the spray. The pulp can be discarded or used around plants as a possible insecticide. Several garden plants contain compounds that are weakly insecticidal. Tomato leaves contain an alkaloid similar to nicotine. Onions, garlic, and turnip leaves have various sulfur-containing compounds, and rhubarb leaves contain oxalic acid. Other plants with chemical compounds that offer potential control of insects are hot pepper fruits, walnut or pignut leaves, and larkspur seeds. Most of the sprays prepared as described from this list of plants would be only mildly insecticidal. All of these plants have materials with potential harmful effects to mammals through direct toxicity, irritation of skin, or allergenic effects.

TABLE 11.2
Some Plant-Derived Dusts for Insect Control

Pyrethrin	Ryania	Quassia
Rotenone	Sabadilla	

Dusts (See Table 11.2)

These plant-derived materials, or botanical insecticides, may be applied as sprays by their suspension in water as well as dusts. These insecticides are broad-spectrum materials, meaning that they have actions against beneficial insects as well as pests. All of them have short-term residual activity on crops, but all of them have mammalian toxicity. Care should be taken when choosing botanical insecticides.

Pyrethrin

Pyrethrin, also called *pyrethrins* or *pyrethrum*, comes from one or more plants of the *Chrysanthemum* genus, called *painted daisies*, *bug-killing daisies*, or *pyrethrum daisies*. The botanical insecticide is available as powdered flowers of these plants or as an extract of the flowers and is referred to generally as *pyrethrum*. Pyrethrum is the most commonly used of the botanical insecticides in the United States. Synthetic forms of pyrethrin are known as *pyrethenoids* and are ingredients in many of the popular house and garden sprays and in some sprays known as *white-fly sprays*. Natural pyrethrum only may stun insects and knock them to the ground. Insects often metabolize pyrethrum and recover. Synthetic forms in spray cans are fortified with a synergist, piperonyl butoxide, which also kills insects. Pyrethrin is a broad-spectrum insecticide that poisons the nervous systems of insects and that has low mammalian toxicity. It is perhaps the least toxic of the botanical insecticides. Commercial preparations of natural pyrethrum are sold as liquid preparations often with rotenone or with insecticidal soaps. Pyola is a broad-spectrum insecticidal spray prepared from pyrethrum and canola oil and is used against adults, larvae, and eggs of insects.

Homemade pyrethrum dusts can be made from powdered, dried, mature flowers of the painted daisy. Extracts can be made with by soaking a cup of packed flowers in 2 fluid ounces of rubbing (70% isopropyl) alcohol. The alcoholic extract can be diluted with water to make a spray. Pyrethrin seems to lose some effectiveness at temperatures above 80°F.

Rotenone

Rotenone is a plant product refined from roots of tropical Asian (Malaysian derris, *Derris* spp.) and South American (Peruvian cubé and Brazilian tembo, *Lonchocarpus* spp.) legumes. Rotenone is a broad-spectrum insecticide with high toxicity to chewing insects (e.g., various beetles) but provides lesser control of sucking insects (e.g., aphids) and of larvae and slugs. Rotenone is a respiratory poison interfering with the electron transport chain in mitochondria. It is quite toxic to mammals, birds, and fish. Rotenone was used by American Indians as fish poison. It readily enters the bloodstream of fish through the gills, and the poisoned fish come to the surface and are caught easily. The fish could be eaten safely because rotenone is not absorbed readily in the gastrointestinal tract. Rotenone is used commonly today in fish management to eradicate unwanted or exotic fish from non-native bodies of water. Care should be taken not to use rotenone where it is likely to enter ponds and other bodies of water. Rotenone has residual effects on crops for about 1 week, so harvests should be scheduled for at least 1 week after rotenone applications. Commercial powdered preparations are 1% or 5% rotenone. The remainder of the mixtures is inert ingredients (carrier). Commercial mixes are sold as these dilute preparations because of the high toxicity of rotenone. The 1% preparations generally are used as dusts, whereas the 5% preparations may be used as dusts or as sprays. The preparations meant to be used as sprays should be wettable powders; otherwise, wetting of the powders meant for dusting is difficult. Rotenone has a shelf life of a year or less, and year-old materials are likely to be ineffective. Therefore, new materials should be purchased yearly, and growers should purchase only enough material for the current season to avoid waste and disposal of the material. Rotenone products, in addition to those marketed with it as the single active ingredient, may be sold as powdered mixes of rotenone, ryania, and pyrethrin. Rotenone (1%) is also available as an emulsifiable concentrate with pyrethrin (0.8%) that can be mixed with water and sprayed. Rotenone is a very effective broad-spectrum insecticide, but because of its toxicity and because some evidence indicates that it causes growth abnormalities in test animals, growers may want to choose other organic insecticides that are safer to use. Rotenone is more toxic to mammals by inhalation than by ingestion. Applicators of rotenone should wear protective clothing and masks and follow all instructions written on the label for the material.

Rotenone has been used to control fleas, lice, mites, and insect larvae on livestock, pets, and poultry. Rotenone formerly was permitted in organic agriculture. It was removed from the list of approved substances of the USDA National Organic Program in about 2005 because of concerns about its safety. Except as a fish poison, its use is being phased out in the United States.

Ryania

Ryania is prepared from the stems and roots of a shrub (*Ryania speciosa*) from South America. It is a broad-spectrum, general-use insecticide with strong effectiveness against the larvae of various moths. It is recommended to control larval pests, such as corn borer, corn earworm (tomato fruitworm), cabbageworm, and codling moth, and for control of thrips, aphids, and several beetles. It has limited use against cabbage maggot, cauliflower worms, or boll weevil. Users should consult the label on the product to ascertain its permitted uses. Ryania is difficult to find in garden and farm stores but is available by mail order through garden supply catalogs. It is available as a single product or as a mixture with pyrethrum and rotenone. It can be applied as a spray or dust. Repeated applications at approximately 10- to 14-day intervals may be needed to provide season-long control of pests. Applications should be ceased several weeks before harvest of produce. Ryania is effective in warm weather and has a shelf life of several years if kept cool and dry. Ryania is a much safer material than rotenone. Its mammalian toxicity is similar to that of pyrethrum. Users should wear protective clothing and masks.

Sabadilla

Sabadilla is made from the seeds of *Schoenocaulon officinale*, which is a lily-like plant native to Venezuela. The seeds contain poisonous alkaloids that are not in other tissues of the plant. Sabadilla is a broad-spectrum insecticide that offers protection against larvae and adults. Usage is most common as a dust, although the powder may be wetted and sprayed. Applications to plants should be about weekly to maintain control of pests. Moisture on the leaves helps the dust to stick to plants. Sabadilla degrades quickly in sunlight, moisture, and air so that it does not persist in the environment. However, it has a long shelf life and seems to increase in potency with age in dark, dry storage. Sabadilla has low mammalian toxicity if ingested, but applicators should avoid inhaling the dust, as it can be irritating to mucous membranes. It is quite toxic to honeybees. Recommended times of use are in the evening after the bees have returned to their hives.

Quassia

Quassia comes from the bark and wood of a tropical American tree (*Quassia amara*) with bitter wood. This species is in the same family as tree of heaven (*Ailanthus altissima*), also with bitter wood. Wood shavings or bark chips of quassia are spread on soil at the base of plants to control insects. An extract of quassia may be made from ground wood or bark with hot water. The extract seems to be a weaker insecticide than the straight bark or wood. Quassia is sold as a medicinal herb and can be purchased in some natural food stores.

Lime-Sulfur

Lime-sulfur dusts can be made with mixed proportions of various limes (agricultural lime, quicklime, or hydrated lime) and sulfur. The ingredients also can be used individually. Quicklime or hydrated lime is a little more effective than agricultural lime alone or with sulfur because of their amorphous nature and reactivity. The dry materials will not burn foliage. Growers will want to take caution against application of these materials to wet foliage. Hydrated lime mixed with water is whitewash, which can be used for insect control on trunks or stems of woody plants.

BIOLOGICAL CONTROL (SEE TABLE 11.3)

Biological control in the context that it is used here refers to the intentional introduction of parasites, predators, or diseases of pests against insects. Enhancement of the effectiveness of beneficial organisms or use of conditions already present in the environment will be discussed under the topic of *natural control*. Biological or natural control procedures often are used in preference to spraying or dusting with chemicals. Spraying or dusting with broad-spectrum insecticides, organic or otherwise, upsets the balance of beneficial insects in relation to their prey. Organic botanical sprays or dusts that are broad-spectrum insecticides kill a wide range of

TABLE 11.3
Biological Agents Including Predators, Parasites, and Diseases for Control of Insects

Insect Predators and Parasites	
Ladybugs	Parasitic wasps
Praying mantids	Mites and spiders
Lacewings	Nematodes
Insect Diseases	
Bacteria	Fungi
Protozoa	

insects including beneficial insects that are effective in biological control. In some cases, pests that survive the spraying or dusting may be unchecked if the beneficial organisms are killed. Outbreaks of pests often occur after application of a broad-spectrum insecticide, because of the lack of competition between pests and their enemies. Use of biological control and especially natural control helps to maintain a balance of predator to prey, such that as the pest population rises in the environment the beneficial organisms increase to keep the pest in control.

Beneficial organisms that a grower can use in biological control include various insects, predatory mites, nematodes, bacteria, fungi, protozoa, and viruses. Garden supply catalogs offer a wide variety of these organisms. Private laboratories also offer organisms for sale directly to customers. Most of these purchased beneficial organisms need special care. Growers must learn about the life cycles and feeding habits of the organisms. Unless the organisms have the prey around when released, the beneficial organisms may die or leave the premises. Many of the organisms are not winter hardy and will have to be reintroduced every year in cold climates. Many of the organisms also are available only in limited supply from private laboratories. Some tend to be expensive. Also growers should consider that biological control may not be as fast as the grower wants, and the temptations to use sprays and dusts must be abated.

Methods of Biological Control

Insects against Insects

Ladybugs

These insects include several species of beetles of somewhat similar appearances. They may be orange bodied with black spots or black bodied with orange spots, ranging from two to twelve spots depending on species. Beetles for sale generally are harvested from the redwood region of California and are orange with twelve spots. Ladybugs are used against eggs, larvae, and adults of small, soft-bodied pests, such as aphids, some scales, and spider mites. Adults and larvae of ladybugs are active predators. Ladybugs are effective in control of pests that are on the undersides of leaves and in the buds and whorls of shoot tips where sprays and dusts may be ineffective because of the difficulty of bringing the insecticide in contact with the pests. The use of ladybugs is controversial partly because of the methods of their harvest for sale but primarily because they do not remain on the property on which they are released. Ladybugs have the tendency to fly away, especially if no prey is available for them to eat when they are released. Hence, some recommendations advise their use only in greenhouses. Dormant ladybugs may be stored in the refrigerator for about 2 weeks if outside conditions are not conducive for their release. Some retailers sell ladybugs for which their dormancy has been broken under laboratory conditions and their initial tendency to fly away is less than that of unconditioned ones. Even the conditioned ones must have prey present if they are to remain around. Some retail catalogs offer a protein-based bait that can be sprayed on foliage with the release of insects. This bait may help to retain the beetles or other released insects, such as lacewings, on the premises. About 500 bugs (0.5 pint) will cover about 1000 sq ft of outdoor or greenhouse space.

Praying Mantids

The mantids will eat any insect that they can catch, large or small, including aphids, leafhoppers, flies, spiders, tent caterpillars, and others. Mantids will eat beneficial insects as well as pests, and their effectiveness in biological control may be low because of this fact. Mantids are bought through the mail by the purchase of egg cases. Three cases will provide enough adults to cover 5000 sq ft of outdoor space. The mantids usually stay around, multiply, and perpetuate themselves. Egg cases usually are shipped from November through May. Cases shipped after May can hatch in the mail, and mantids will eat other mantids.

Lacewings

The green lacewing is said to be the best all-purpose beneficial insect for sale. A problem exists, however, with poor hatching of eggs that are shipped or poor survival of the larvae. Surviving lacewings are very effective in control of pests. The larvae consume aphids, some scales, mealybugs, whiteflies, and eggs and larvae of many pests. Living on nectar, adult lacewings do not eat pests. Several releases of lacewings may be needed to get a population established. Each release should be about 1000 to 5000 eggs per 1000 sq ft of outdoor or greenhouse space.

Parasitic Wasps

These wasps are small and, having wing spreads of 1/16 inch or less, are harmless to humans. The *trichogramma wasp* is a parasite of insect eggs of worm (larval) pests, such as tomato hornworm, cabbage looper, corn earworm (tomato fruitworm; cotton bollworm), corn borer, and codling moth. The action of the wasp is by parasitic action of the larvae, which enter and consume the eggs. Releases of wasps should be in large numbers at one time or in three releases at 2-week intervals. Typical releases are 5000 eggs per fruit tree or per 100 sq ft at biweekly intervals. Eggs are sold by mail order and are shipped at the appropriate time of the year to avoid hatching of the wasps in the mail and their subsequent loss by their escaping when the packages are opened. The whitefly parasite (*Encarsia formosa*) is another wasp that is useful for whitefly control in greenhouses. *Encarsia formosa* is effective if released when whitefly populations are low, for the wasps alone cannot control populations of whiteflies that are large. Control is more unlikely during the winter months than during warm sunny weather. The sweet potato whitefly is difficult to control by the wasp. Several releases of wasps are needed for control of any kind of whitefly in greenhouses. This wasp is available from insect-rearing laboratories and retail catalogs. Other wasps in the wild attach to pests such as tomato hornworm. The white bodies that are seen on hornworms are wasp eggs. The larvae hatching from these eggs enter the bodies and parasitize hornworms. Hornworms that have these eggs attached are doomed, and growers should let the parasite take its course so as to not break up its life cycle, although this practice is difficult as growers note that the hornworm continues to feed with the attached eggs.

Predatory Mites and Spiders

Mites are arachnids, eight-legged animals (arthropods) related to spiders. Predatory mites are less than 1 mm (1/25th inch) across. Their prey is other mites, thrips larvae, fungus gnats, and a few other pests. They may not control adults of insects, such as thrips. Predatory mites can be used to help in pest control in greenhouses or in gardens, vineyards, and orchards. Predatory mites may exist naturally in the environment. Spraying may kill off the natural population leading to a proliferation of pests that were controlled by the predators. Predatory mites are available for sale from insect-rearing laboratories. Several species are offered. Common genera of predatory mites are *Phytoseiulus* (spider mite control on strawberry and in cool greenhouses), *Amblyseius* (use in hot greenhouses), and *Metaseiulus* (use in orchard and berry patches). Populations of mites released should be 2000 to 5000 per 1000 sq ft of greenhouse or garden space or 100 to 1000 per fruit tree. The larger numbers at the top of these ranges should be released to control an outbreak of pests during a current season. Fewer mites, numbers at the lower end of the preceding ranges, can be released if the objective is to build up their population in the environment.

Spiders are eight-legged arthropods, generally larger than mites. Spiders come in many different colors, shapes, and sizes. Insects are their main diets. Insects are usually caught in the spiders' webs; very few spiders are large and fast enough to chase down their prey. Most spiders are nonpoisonous to humans. The black widow and the brown recluse are well-known spiders that are poisonous to humans. Refraining from spraying with insecticides and miticides is the best way of encouraging spiders to populate the garden.

Beneficial Nematodes

Disease-causing nematodes are microscopic worms and are associated most often with parasitic plant diseases (root-knot nematode, for example). Beneficial nematodes are large, parasitic organisms that attack larval stages of soil-borne pests, such as Japanese beetle grubs, June beetles, army worm, cabbage root maggot, wireworm, cutworm, and root weevil. These pests have their larval stages in the soil or the larvae crawl across the surface of the soil where they come in contact with the beneficial nematodes, which reside shallowly in the soil or its surface. Species of beneficial nematodes are *Steinernema carpocapsae* (abbreviated Sc) and *Heterorhabditis heliothidis* (abbreviated Hb). Beneficial nematodes are fast acting in the soil. They apparently search for their hosts, but their mobility is limited in undisturbed, agricultural soils. The nematodes enter the hosts through openings such as the mouth of the larvae. Inside the larvae, the nematodes release a bacterium (*Xenorhabdus* sp.), which kills the larvae. The nematodes consume the dead host and then leave it for another host. In the soil, nematodes have a protective cuticle and can survive without a host for a year if the soil moisture and temperatures are favorable for their survival. Nematodes do best in warm, moist conditions. Outdoor conditions may limit their effectiveness in pest control. In the winter, they burrow into the soil and hibernate. In the spring, they move closer to the surface, but their return lags that of the pests. For early control of pests in the spring, growers must reintroduce nematodes or use another means of control until the originally introduced soil-borne nematodes are effective again.

Other Predators

Many other predatory insects are available for biological control of pests. Some of the predators and parasites that are not discussed here are as follows (control in parentheses): aphid midges, aphid parasites, soldier bug (larvae of bean beetle, potato beetle, and sawfly; army worm; tent caterpillars and other caterpillars), pirate bugs (various insect larvae and eggs), *Cryptolaemus* (mealy bug), and scale predatory beetles.

Insect Diseases

Several microbial pesticides are available for insect control. These pesticides are highly host specific, and users must have considerable knowledge of the pesticides and targeted insects. The microorganisms that bring about control are specific for kinds and stage of development of insects.

Bacillus thuringiensis

Bacillus thuringiensis is perhaps the most widely used microbial pesticide, which is known by its common name Bt. It controls larval pests, such as the cabbage looper, gypsy moth larvae, codling moth larvae, tomato hornworm, spruce budworm, tomato budworm, corn borer, and larvae of other moths and butterflies. Some wormlike pests, such as slugs, are not affected. Many varieties of *B. thuringiensis* have been identified. The most common variety used as a pesticide is *B. t. kurstaki*, which sometimes is called Btk. Btk is applied directly to foliage of plants. Other varieties are *B. t. israelensis* (Bti), which is marketed for mosquito, blackfly, and fungus gnat control. Bti is applied to standing water or as a soil drench. Floating brickettes of Bti are available. *B. t. san diego* (Btsd) is marketed for control of various leaf-eating beetles and is applied on foliage. Each of these varieties is sold under several trade names, and the user must become familiar with each brand and its use by reading thoroughly the descriptions of the products before purchasing them. Bt lasts for 1 to 4 days after spraying and must be reapplied for continued control. Some encapsulated preparations are available to give prolonged control. The Bt produces a crystalline proteinaceous toxin that poisons various larvae. After Bt is applied to plants, the larval pests ingest it as they feed on foliage. The larvae soon stop feeding after ingestion of the pesticide although they may live for several more days. Eventually the larvae discolor, drop to the ground, and die Bt is available in granules, dusts, wettable powders, and liquid concentrates. The dry forms have a shelf life of several years in a cool, dark, dry storage place. The liquid concentrates have a shelf life of 1 year. Preparations mixed for pest control should be used in a few days. Commercial appetite stimulants are available to encourage the larvae to eat more foliage and thereby to consume more Bt. The stimulants also can result in more damage to leaves by the current or following infestation of larvae. A few confirmed cases of development of resistance to Bt have been documented.

Milky Spore Disease

Bacillus popilliae and *Bacillus lentimorbus* are bacterial pesticides for control of soil-borne larvae (grubs) of Japanese beetles and June beetles. These larvae feed on grass roots and are food for moles, the burrowing of which does even further damage to lawns. As they bore through the soil and eat on grass roots, the larvae ingest the bacterial spores of milky spore disease. The bacteria multiply in and fill the larvae with a milky concentrate of bacteria and spores, which survive in the soil after the larvae die. Earthworms are not harmed by milky spore disease. Milky spore disease is available as dusts or granular powders, both of which should be applied to unfrozen ground. The dusts are applied as a spoonful every 3 to 5 feet across the lawn, and the granular forms are broadcasted. Control may be permanent. In warm climates, the disease organisms multiply as they feed on grubs, and burrowing grubs help to inoculate the soil. In northerly climates, annual applications of milky spore disease may be required. Adult beetles are not controlled, and control of larvae is limited to the treated areas.

Fungal Diseases

Insects that spend part of their life in soil may by controlled by the fungus *Beauvaria bassiana*, which occurs naturally in soils. This organism kills a few insects each year in most gardens and fields. Control is not limited to soil-borne insects. Some pests that are affected are Colorado potato beetles, thrips, whiteflies, mealy bugs, and various caterpillar pests. Population of the disease rises in response to outbreaks of serious infestations of pests, but often the damage by the pest is severe before native populations of the disease bring the pest into control. Commercial products of the disease are not now available in the United States. These materials are spores that are sprayed on crops. The fungus *Verticillium lecanii* is a biocontrol agent used in Europe against aphids, mites, and mealybugs in greenhouses. Fungal diseases of insects are being researched for use in the United State. Until these organisms are commercially available, growers should encourage natural populations in the garden by limiting uses of fungicides.

Protozoa

Protozoa are single-celled organisms, classified as animals in taxonomy that divides life simply into animal and plant kingdoms. The parasite *Nosema locustae* infects grasshoppers and crickets. Once ingested, usually on bait, the organism multiplies in the guts of the pests and slowly kills or makes the pests very sick and inactive. Successive generations of grasshoppers are infected by the preceding one. Control of infected insects is normally slow, sometimes requiring only hours but more commonly requiring days or weeks. Another major limitation in the use of *Nosema locustae* is that grasshoppers are mobile insects, and adults from untreated areas will migrate into areas in which populations have been controlled; hence, control is not apparent. The product is inexpensive so that sizable areas can be treated economically, but a community effort may be needed for effective control. Repeated applications likely are necessary. It is a registered, commercially available microbial agent for grasshopper control on rangeland.

CULTURAL CONTROL

Cultural control involves practices that growers employ during crop production to protect plants from insects (Table 11.4). These practices include anything that growers can do to encourage development of healthy plants, along with use of barriers and traps, timing and placement of crops, and other procedures to avoid introduction or to impede growth of insects. In this section, practices are divided into those that involve crop management and those that involve physical controls.

TABLE 11.4
Cultural Procedures for Control of Insects

	Crop Management	
Sanitation	Timing	Crop rotation
Tillage	Plant combinations	
	Physical Controls	
Barriers		
Row covers	Paints	Mulches
Disks and collars	Bands	
Traps		
Sticky substances	Water	Baits and scents
Trap plants	Ditches	Lights
Manual		
Handpicking	Vacuuming	Pruning
Washing		
	Miscellaneous	
Insect-resistant varieties	Plant nutrition	

Crop Management

Sanitation

Insects, such as pea weevils, cucumber beetles, squash bugs, and codling moths, may live in crop residues and trash that are left around cropped areas. Crop residues should be turned into the ground or removed from the site. The site must be kept clean of debris. The cropped area should be clear of old baskets, sacks, crop residues, dropped fruits and vegetables, and other trash. Plants with heavy insect infestations should be pulled and discarded.

Tillage

Tillage helps to break the life cycle of insects. Insects that are born in crop residues may not be able to survive in residues that are incorporated into the ground. Tillage also physically destroys the crop residues and destroys shelters in which insects are residing in the soil. Tillage to prepare the seedbed disturbs insects at their most vulnerable stage, exposing grubs and adults to the harsh environment of the atmosphere where they may freeze, overheat, or desiccate or be eaten by birds. Tillage may separate insects from their food sources by the destruction and distribution of residues in the soil. Tillage controls weeds in which insects may be living before planting. After-harvest tillage helps to control corn borers, tomato hornworms, pea weevils, squash bugs, and other pests that are in crop residues. Tillage with a moldboard plow buries insects deeply. A rotary tiller is most destructive of organic residues but provides only shallow disturbance. Use of hand tools is much gentler on insects and residues than use of power-driven implements. Employment of the practice of no-till agriculture lacks the benefits that are received from insect control from or during tillage. However, decreased tillage favors development of soil-borne organisms, including insects, their predators, and earthworms.

Soil-borne pests may emerge earlier in tilled land than in untilled land, because tilled land warms faster. Timing of planting in relation to emergence of pests is important. Early emergence of insects in a newly planted area can result in serious crop losses, whereas early emergence of the pests before the crop is planted may allow for the pests to leave or to starve.

Timing

Timing of planting and harvesting can be effective in avoiding damage from insects. Early plantings of crops may withstand insect infestations that arrive after the crops have developed sufficient size and vigor so that the damage inflicted on the crops is small relative to that suffered by young, late-planted crops. Early plantings of onions are more able to resist the onion maggot than plants that are young when the maggot emerges. The onion maggot emerges first at about the time that dandelions bloom. A large-stemmed squash plant has more resistance to borers than a seedling when the borer emerges.

However, a delay in planting may prevent a crop from being exposed to insects. Maggots are a serious pest on brassicas in the north. Cabbage and other brassicas planted after the egg-laying fly has come and gone—usually in early June—will not be infested. If the timing in June is uncertain, these crops can be planted in early July for a fall harvest with good certainty of avoiding the maggots. Planting of radishes before April 1 or after May 1 may permit avoidance of their infestation by maggots. Fly-free dates are consulted for timing of planting to protect wheat against the Hessian fly in the Great Plains. Degree-days (cumulative time above a certain temperature, such as 50°F) can be used to schedule planting, harvesting, or control measures. Phenological signals, such as blooming of trees, shrubs, and weeds in relation with emergence of insects, can be correlated with emergence of insects and used to time planting, harvesting, and control measures. Planting seeds, especially bean and corn seeds, in warm soil results in their rapid germination and helps to prevent their destruction by seed maggots.

Crop Rotation

Crop rotation can be an effective means of insect control. Insects often are quite host specific, and changing crops may starve out certain pests or reduce their numbers. Problems arise when an insect attacks a broad spectrum of plants. Flea beetles are problems with tomatoes and potatoes. The tomato fruitworm, tomato budworm, corn earworm, and cotton bollworm are all the same pest, known by different names depending on the plant organ or species attacked. Crop rotation becomes complex if a large variety of crops are grown in a limited space such as a home garden. Gardeners must become fully knowledgeable of taxonomy of crops and insects. Also, consideration must be given to the allelopathic or other growth-limiting effects that crop residues have on subsequent crops. In most cases, only slightly restricted crop growth results from allelopathic effects. Fertilization will overcome effects that may be imparted by coarse, nitrogen-poor residues.

Plant Combinations

Use of plant combinations in organic gardening is known as *companion planting* (see Chapter 13). The concept of pest control behind companion planting is that a diversified grouping of plants reduces insect problems relative to those that occur in a grouping of only one kind of plant. The plant-to-pest relationship in diverse plantings may be expressed in several ways. The suggestion that a diverse planting confuses and thwarts insects is not well supported by evidence. Plantings in which two or more crops have been alternated in rows or hills often have lower individual and total yields than pure plantings and still may not show any evidence of pest control. However, some plants act as attractants to bring beneficial insects into the garden. These plants can be spaced among the crop plants or planted in borders so as to not interfere with the normal spacing of the crop. Plants of the carrot family, aster family, and mint family are suggested as plants to attract beneficial insects into the garden.

Other plants (see the section titled "Trap Plants") may be used to attract pests away from the garden, where the pests can be destroyed. Some growers suggest solanaceous weeds (nightshade family) to attract Colorado potato beetles from crops; others suggest corn to attract earworm from tomato (the tomato fruit worm or bud worm is the same organism as the corn earworm). The practice of using plants to divert pests from a garden or field is relatively untested, and substantial experimentation on the part of the grower may be needed. Also, the use of plants to repel insects has little scientific evidence in its support. The beneficial effects of marigolds are attributed to the action of their roots on control of soil-borne nematodes. Marigolds probably have no above-ground action in insect control. Onions and garlic, which have insecticidal compounds in their tissues, probably have no effect on deterring insects in the vicinity of these plants. Little beneficial effect on disease control other than that imparted by control of insect vectors is attributed to companion planting.

Physical Controls

These control measures involve barriers and traps to keep insects from crops. Some methods are listed here, and innovative growers should be able to expand this list.

Barriers

Barriers keep pests from plants. Barriers should be in place before the insects arrive at the crops; otherwise, control measures must be employed on the crop directly. Some barriers to control insects are included in the following discussion.

Row Covers

Row covers (Figure 11.1) that are placed over crops to offer protection against frost also provide barriers against insects that fly or crawl toward crops. Protection is offered against aphids, various beetles, maggots, caterpillars, and leafhoppers. No protection will be provided against soil-borne insects. Mulches may be needed for that purpose. For dual-purpose use, clear *polyethylene plastic* gives the best warming conditions among row covers. Polyethylene must be supported by hoops to keep it from crushing plants, and also it must be ventilated in some manner to keep temperatures from rising too high on warm days. Usually polyethylene is removed from the rows after danger of frost has passed and when the daytime temperatures outside the plastic are so high as to give a danger of excessive heating under the plastic. *Floating row covers* are made of fabric, such as spun-bonded polyester or polypropylene. This fabric is light weight and does not have to be supported over the plants. Water penetrates through the fabric so that rainfall and irrigation can reach the crops. Although floating row covers are used mainly to extend the growing

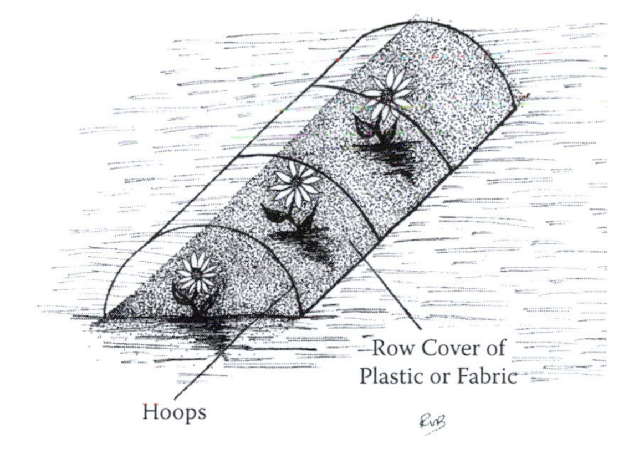

FIGURE 11.1 Placement of covers over rows of crops.

season, either in the spring or fall, they also provide good protection against insects. Floating row covers provide good air circulation around crops, but frost protection may not be as good as that provided by polyethylene covers. Weed control is essential under row covers, and covers may have to be removed for this purpose unless a weed-controlling mulch, such as black plastic, is laid down before the covers are in place. Covers will have to be removed when cross-pollinated plants begin flowering. The fabric covers can be cleaned by laundering and used again providing too much physical deterioration has not occurred. Holes can be patched with tape or by sewing. The spun-bonded polyester covers are cheap enough that recycling is not practiced generally. Other materials that give protective covers against insects are cheesecloth, shade cloth, wire cages, metal or plastic screen wire, and hotcaps.

Mulches

Mulches provide a barrier between the soil and shoots of plants. Mulches are a barrier for insects that are headed for the ground from external sites. Reflective aluminized or white mulches help to deter flying insects from landing in the garden or field. Warming from plastic mulches may deter pests that cannot tolerate heat.

Bands

Bands (Figure 11.2) of sticky or wooly matter act as barriers to pests that crawl on trunks of trees, shrubs, and vines. Ants, gypsy moth caterpillars, codling moths, some beetles, snails, and slugs may make daily trips up and down tree trunks. Sticky coatings may be applied directly to the plant as long as the coating is not phytotoxic, but applications on a removable wrap are usually better. The wrap helps to prevent damage to the bark directly from the sticky material or from fungi that may grow in the material. Bands should be at least 2 inches wide and be placed about 18 to 24 inches off the ground. Bands should be removed about once a week to get rid of insects that have accumulated in the bands. Bands filled with pests can be removed more easily with wraps than with direct sticky applications to the bark. Commercial materials are available to make the removable wraps to which sticky materials can be applied. Wooly bands are also commercially available. Some commercial bands are slippery adhesive tape coated with silicone over which caterpillars are unable to cross. Wraps can be made from cardboard or corrugated pasteboard, which are taped or stapled in place. Some insects may pupate in the corrugated pasteboard, which

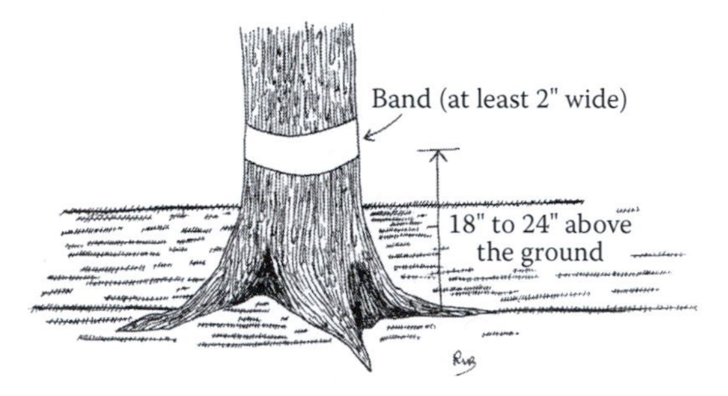

FIGURE 11.2 Placement of bands for control of crawling insects on tree trunks.

can be removed and destroyed. Aluminum foil also can be used to make bands which are too slick for caterpillars to cross or which can be coated with sticky material. Burlap, canvas, flannel, or other cloth strips can be tied with cords around trunks to create a zone that is a barrier or trap. These bands may require some sealing with sticky material at the edges to prevent insects from crawling under them. Bands may be impregnated with predators or insecticides (see sprays and dusts above) to give further control against crawling insects.

The sticky material should not be petroleum-based lubricating grease. In hot weather, the grease may melt, run down and under the band, and permeate into the bark, effectively girdling the tree. Commercial nonphytotoxic gums and bug-tangling preparations are available. These sticky materials are made from natural gum resins, waxes, and oils from plants. They are provided as sprays and as ointments in caulking gun cartridges, tubes, and tubs.

Painting

A protective barrier against boring insects can be painted on trunks with whitewash (slaked lime in water) or with latex paint (diluted with equal parts of water if tested before use). The barrier should begin about 1 inch below the soil line and extend 2 to 3 feet above the ground. This barrier also protects trees against sunburn or bark cracking and against fungi. Tape is available for wrapping trees to give about the same effect as whitewashing or painting.

Seedling Protectors

Egg-laying flies, such as the cabbage maggot fly, deposit their eggs on the soil at the base of the stems of vegetable seedlings. With disks, used as a barrier, cabbage, cauliflower, broccoli, and other cole crops can be protected from the maggot flies. Disks (Figure 11.3) of tar paper or cardboard (not corrugated pasteboard) can be cut 3 or 4 inches in diameter or larger up to 8 inches, slitted, and placed over the base of young plants. The disks can be round or square, but too small disks may curl and give no protection. The disks provide a barrier that keeps the flies away from the soil at the base of the plant. Tar paper makes the best disks, and although some repelling action on the flies is attributed to the tar paper, its durability is its major strength. Kraft paper is too thin to use. The disks must be kept clean of soil and mulch to be effective. The disks also must be snug fitting around the stem of the seedling; otherwise, enough room is left for the flies to lay eggs. Growth of the stem will push the slit of the disk outward with no harm to the plants as they grow. Do not use stiff plastic disks, such as margarine cup lids, for these will cut the stems. Do not use soil to hold the disks in place, since the soil on the disk can be sufficient for the flies to lay eggs and to parasitize the plants.

Collars (Figure 11.4), called cutworm collars, can be placed around the base of seedlings to protect them against cutworms and other crawling pests. Paper strips about 2 to 3 inches wide can be wrapped around the base of the stem. The strip should extend into the ground at least one-half inch. Cardboard tubes from paper towels or toilet paper and tin cans with tops and bottoms removed also make cutworm collars. Other suggested barriers against cutworms are toothpicks, match sticks, or nails placed at the base of the plant to prevent the cutworm from wrapping around the seedling.

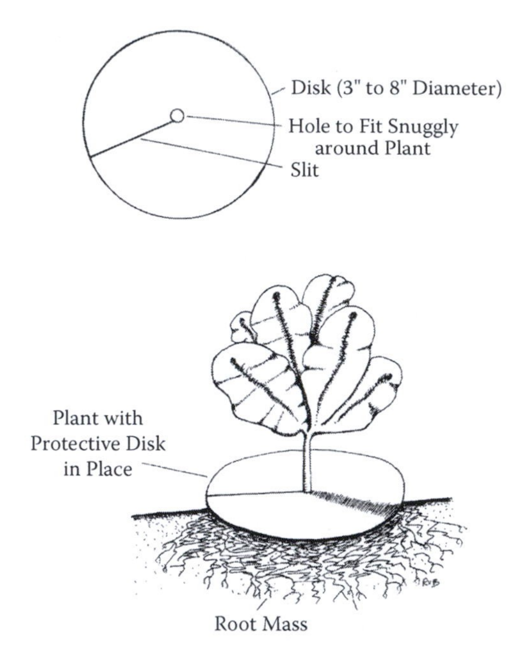

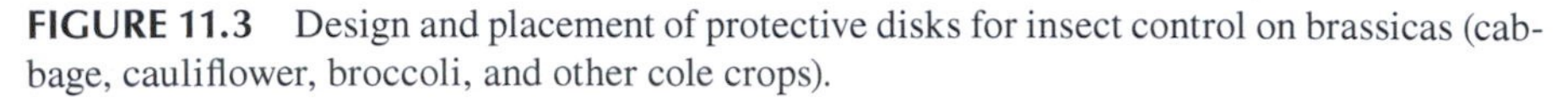
FIGURE 11.3 Design and placement of protective disks for insect control on brassicas (cabbage, cauliflower, broccoli, and other cole crops).

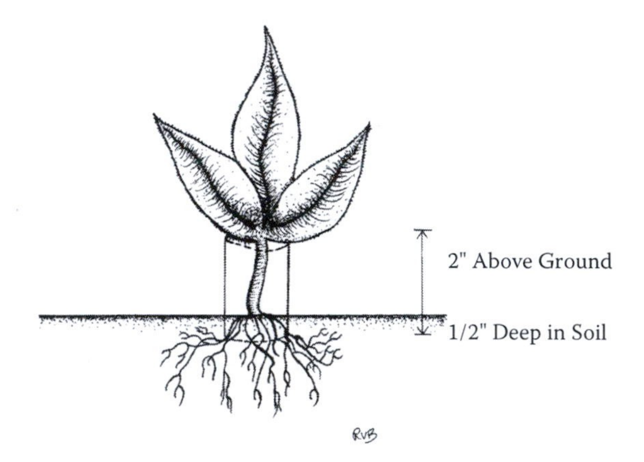

FIGURE 11.4 Placement of paper collar for cutwork control with seedlings.

Traps

Sticky Traps

Paper, wood, or other materials can be sprayed with sticky materials such as those described in the section titled "Bands." These traps, concentrated in small areas, can be used to reduce the population of insects in the area of the plants or as devices from which counts can be made to assess the population of insects in the area. The traps can be of various sizes and shapes, since no particular configuration has been shown to be better than another in most cases. Color of trap may be important. Often, the traps are painted yellow, blue, white, or red to attract flying insects. Yellow is a popularly used color to attract various insects. Yellow is the most common color used to catch whiteflies and other pests in greenhouses. Blue is used to trap thrips, and white is recommended for flea beetles and tarnish bugs. Red is used to mimic fruit colors and to attract apple maggot flies. Yellow is also used in orchards to trap cherry fruit flies. In orchards, shape of traps may be important, with fruit-like shapes being used commonly and made from objects such as rubber balls or bottoms of plastic soda bottles. Sticky traps are placed at plant height or hung in trees. Hanging by wires is a convenient method of inserting the traps among plants. Traps should be kept away from foliage and out of the way of workers. Decisions on the necessity of other control practices can be made based on the estimated or correlated populations of insects made from counts of the number of pests trapped over a given time.

Baited or Scented Traps

In small areas, traps with baits may lure pests away from susceptible crops. Traps can be made from vegetables or fruits, such as lettuce, carrots, potatoes, onions, oranges, grapefruits, and apples. These fruits or vegetables can be placed in containers to trap insects or may act as the traps without containers. The insects are attracted to the traps, which are discarded with the trapped pests. Vegetable and fruit traps work better for night-feeding insects than for those that feed during the day. Mature cull onions may be planted near or before sets or plants are planted for the main crop. The mature onions will sprout and will be attractive to pests such as the onion maggot. The infested plants can be pulled and discarded. Beer has been used to trap slugs; some beer is placed in a container into which the slugs are attracted and drowned.

Baits may be used in conjunction with sticky traps or some other form of trap. Various foods (molasses, sugar, egg albumin, yeast, casein) or scents (such as those of fruits, flowers, ammonia, and female insects) are used often as baits. Many of the commercially available traps for professional scouts and for home gardeners include some attractant to lure the pests to traps. *Pheromone traps* are a sort of baited trap for male insects. Female sex pheromones help males find mates, and traps baited with these scents help to restrict the propagation of future generations of certain pests. Pheromones are used to lure insects such as codling moths, cabbage loopers, cherry fruit flies, corn earworms, leafrollers, and Japanese beetles into traps. These traps are readily available for sale in retail stores and catalogs. New, scented traps may need to be set out several times during the season, because the effect of the scent wears off with time. Traps work best if the area in which they are placed is isolated from other sources of the pest being trapped. When traps become full, they have to be disposed of or cleaned.

Trap Plants

Plants that attract certain insects may be planted as borders around crops or among the crops. The concept is that the trap plants will attract insects preferentially and keep the pests from the crop plants. The infested trap plants are then pulled and destroyed with the insects. This kind of control works best with insects that produce only one or only a few generations during the growing season.

TABLE 11.5
Some Plants That May Be Used to Attract and Trap Specific Insects

Trap Crop	Trapped Pest
Dill	Tomato hornworm
Mustard	Cabbage loopers
Nasturtium	Aphids, cucumber beetles, flea beetles, squash bugs
Radish	Cabbage maggot
Soybeans	Bean beetles
Dead nettle	Potato beetles

Drawbacks are that the trap crops use space that could be used for the main crop and that the trap crop may attract insects into the area and lead to an increase in population of pests, particularly if care is not taken in managing the trap plants. Care should be taken to ensure that the trapped insects stay on the trap plants when they are pulled. Some covering of the trap plants during their removal may be necessary to retain the pests. The trap crop should be planted earlier than the main crop to ensure that the trap crop is growing and is more attractive to the pests than the main crop. Trap plants can be used in greenhouses or outdoors. It should be noted that in some reports, trap plants are suggested as repelling the same pests that other reports suggest that they attract. So one should not take these suggestions as fact without some experimentation and close monitoring of their effectiveness. Some suggested trap plants and pests controlled are listed in Table 11.5.

Gardeners should watch for weeds and wild relatives of cultivated crops. Weeds or wild plants may be trap plants, or they may serve as hosts and attract pests into the garden. In the latter case, the wild plants should be removed.

Water Traps

Water traps can be set out in basins that are painted yellow to attract winged aphids and cabbage maggot flies. Soap or detergent added to the water breaks its surface tension and helps to sink and drown the insects. The trap should be exposed and should be elevated if the crops are tall growing.

Ditch Traps

Many crawling insects can be stopped by digging ditches or trenches around the garden. Dusty trenches may prevent migration of chinch bugs and armyworms, but these trenches may need to be up to 10 feet wide. Narrower trenches coated with sticky material can have the same effect. The trapped insects can be left to dry out or be buried in the trenches.

Light Traps

Night-flying insects are attracted to lights. Ultraviolet (black lights), blue, green, yellow, and ordinary incandescent lamps attract insects. Red lights are not usually attractive to insects. Insects are more attracted to ultraviolet, blue, and green lights than to yellow lights. High intensities of light are more attractive than low intensities. Lamps may be employed in various ways to trap insects. Lamps may be hung over pails or basins of water or ponds in which insects are drowned or eaten by fish. They may be used to attract insects into areas with birds or domesticated fowl, which eat the insects. Many mechanical variations in combinations of traps and lights are available. In some devices, suction is employed with a fan, which draws insects into a container where they are collected and destroyed. Bait may be used with these traps with lights or without lights. Killed insects may be allowed to drop to the ground or into water where they are eaten by birds or fish. Electric grids of parallel wires are used in electrocuting light traps. Ultraviolet lights or bait are used to lure insects into the wires. These devices can be used inside or outside to protect crops, livestock, or humans from insects. Insects also can be lured by light or bait through funnels and into bottles. The funnel directs the flight of the insects into the bottles, and they have difficulty finding their way from the bottle through the restricted end of the funnel and die in captivity. The trapped insects can be destroyed or saved for collections. All kinds of night-flying insects can be attracted by lamps. Attracted ones include noxious ones such as houseflies, mosquitoes, and gnats and economically important ones such as codling moths and moths of corn earworms, corn borers, armyworms, bollworms, and tobacco and tomato hornworms. Advantages of these traps include their capacity to be automated and their noninjurious effects on higher organisms. Disadvantages include the necessity of cleaning the lamps and disposing of the dead insects and the expense of purchasing and operating the lights. Some criticism of their use has been that the light traps attract more insects than they kill and actually increase the population of the pests; other criticisms are that the portion of the population of pests killed is insignificant and that the traps kill beneficial insects. Many beneficial insects, including bees and wasps, do not fly at night and would not be harmed by the light-equipped traps, but other beneficial insects, such as lacewings and aphid midges, may be attracted and killed by the lamps.

Other Traps

Although good garden sanitation encourages cleaning the garden of boards, sacks, paper, and dropped vegetables and fruits, leaving these materials in the garden may provide traps for earwigs, slugs, and snails.

Manual and Mechanical Control

These controls involve some hand-operated means to reduce populations of insects by removing them from crops.

Handpicking

Hand removal of slow-moving insects is a possible control when their populations are low. A lot of patience and monitoring are necessary to make this process a success. Care should be taken to ensure that insects to be handpicked are nonstinging and nonblistering. Some people may prefer to wear gloves or to use tweezers or tongs. Tomato hornworm, potato beetle, snails, slugs, and some caterpillars can be handpicked. The removed insects can be killed by placing them in jars or other containers to die; placing them in alcohol, oil, or acid; burning, freezing, or boiling them; feeding them to birds; or flushing them down the drain. Strong infestations can be scraped off plants. Others can be daubed with alcohol, although this process is tedious.

Pruning

Infested portions of plants can be removed by pruning. Aphids, tent caterpillars, webworms, leafminers, bagworms, and various egg masses can be removed, with their spreading to other parts of the plant being avoided. The pruned materials with attached insects or eggs must be destroyed by some means. Care must be taken that during pruning diseases are not spread by the pruning tools. Tools should be disinfected by bleach, alcohol, or detergent after pruning for the day and in some cases between pruning of plants.

Vacuuming

Insects can be sucked off plants by portable or tractor-mounted vacuums. Vacuuming should be directed to the noted sites of infestation or to the tops of the plants. Some machines have been designed for vacuuming of potato beetles in fields. Large quantities of insects may be collected. After they are killed, they may be added to compost piles or used as fertilizer. Some machines have rotating blades used in association with vacuuming. The blades chop insects and infested plant material to pulp.

Water Spraying

Although mentioned earlier in this chapter, the physical action of a strong spray of water will wash insects from bark and crevices and knock them off plants and to the ground. In the process of removing the insects, the spray will damage them by breaking off legs, wings, hairs, and antennae so that they cannot resume feeding.

Insect-Resistant Varieties

Progress in development of insect-resistant crop varieties has not been as dramatic as that in development of disease-resistant varieties, but the situation is hopeful. Development of resistance through plant breeding and selection has been largely through physical changes in plants. Hairiness of leaves gives plants some resistance to insects. Leaf hairs are multicellular growths on leaf surfaces in contrast to root hairs, which are unicellular extensions of epidermal cells. Leaf hairs give a surface on which insects do not like to feed or lay eggs, and the hairs have some toxic effects on insects. Insects on hairy leaves often become lethargic and stop eating. Leaf hairs on tomato appear to exude droplets of fluid that are toxic or annoying to insects in addition to the physical conditions that are undesirable for insects to feed and multiply. Resistance in barley to the cereal leaf beetle was imparted by breeding for hairiness of leaves. Potential for incorporating resistance of tomatoes to whiteflies is in the hairy leaves of wild tomatoes. Long husks of corn give a barrier against entry of the corn earworm. Some husks fail to cover the entire ear of corn, and corn with husks that extend well beyond the tip of the ear have resistance against entry of insects, although harvesting and handling of these ears may be difficult. Leaf angles in onion and daylily can affect resistance to thrips. Plants with wide leaf angles with loose morphology in the leaves give less protection than plants that have narrow angles with tight leaf morphology. Tight leaf sheaths are suggested as giving sorghum some resistance against chinch bugs. Beans or peas with tough pods may be more difficult for curculios to penetrate. Some plants contain toxins in their foliage or other organs to impart chemical resistance to insects. Usually, however, some insect is adapted to the chemical components and is a pest. Plants should always be given the proper environmental conditions for their resistance to be expressed. Plant stresses from drought, wet soils, or improper fertilization may cause insect resistance to be negated and render crops susceptible to insects even though these crops are usually resistant in their natural habitat or under good growing conditions.

Plant Nutrition

Plant nutrition is an important factor in resistance of plants to insects. Well-nourished plants will be strong, large, and vigorous and more able to withstand insect damage than weak, spindly, nutrient-deficient plants. Insects may be attracted to the colors of malnourished plants. Excessively fertilized plants, especially those receiving liberal nitrogen fertilization, often are soft and succulent and are excellent material on which insects can feed.

12 Plant Diseases

CAUSES OF DISEASES

A general definition of plant diseases is that they are disturbances that affect the normal function of a plant. These disturbances commonly are divided into *nonparasitic diseases* and *parasitic diseases*. Nonparasitic diseases are referred to also as *noninfectious diseases* and are caused by environmental deficiencies or disturbances. Nutrient deficiency, cracking of fruits, damage from salinity, air pollution, repeated mechanical damage, and unfavorable light or temperature are examples of nonparasitic diseases. Parasitic diseases are infectious and are caused by organisms or by viruses. Infectious diseases are those that most people envision when plant diseases are mentioned. Principal parasitic organisms that cause infectious diseases are various *fungi* and *bacteria*. Fungi are widespread, multicellular, nongreen organisms. Several thousand species of fungi, among about 100,000 total fungal species, cause plants diseases. Fungal diseases are known as *damping-off, rots, wilts, blights, rusts, mildews, smuts, scab, molds*, and other names. Bacteria are virtually everywhere. A few hundred species cause plant diseases. Some common names for some bacterial diseases are *spots, wilts, rots, blights, galls*, and *cankers*. *Mycoplasmas* are unicellular organisms, classed as bacteria but having no cell walls and being smaller than common bacteria. They are like masses of protoplasm surrounded by an outer membrane. Mycoplasma cause several diseases known as *yellows* (carrot yellows, aster yellows). At one time, yellows diseases were attributed to viruses. Occasionally and usually in tropical regions, *algae* and *protozoa* cause plant diseases. Under taxonomic systems that divide living organisms into plant and animal kingdoms, some algae that might cause diseases are classed as unicellular plants having pigments of various kinds of chlorophyll. Under this taxonomic system, protozoa are unicellular animals. Mistletoe and dodder are *flowering plants* that are parasitic on other plants and thus may be classified as diseases. Infectious *nematodes* are roundworms that produce diseases on roots, leaves, stems, bulbs, and flowers. Soil-dwelling nematodes are the most common disease-producing species. *Viruses* are nonliving, noncellular matter that are composed of a core of nucleic acids and a sheath of protein. Viruses are only 1/100 to 1/1000 of the size of bacteria and may infect bacteria. Viral plant diseases cause malformations known as *mosaics, ring spots, rosetting, curling, puckering*, and other names.

Identification of plant diseases is difficult. A grower may not be able to distinguish a nonparasitic disease from a parasitic disease. Infections with parasitic organisms may follow weakening of plants by a nonparasitic disorder. Bacterial and fungal diseases are difficult to discern by untrained people. Therefore, it is essential that expert advice be sought if a malady strikes a crop. Agricultural colleges and extension services usually have resident faculty and agents who are experts on diagnosis of plant diseases. Growers should contact these personnel to obtain information on handling and shipping of plant samples to the trained experts for diagnosis.

SPREADING OF DISEASES

MEANS OF SPREADING

Infestation occurs when the inoculum of a disease-causing organism or virus comes in contact with the plant. The inoculum may include a virus particle, a bacterium, a fungal mycelium, a bacterial or fungal spore, a nematode, or a seed or portion of a parasitic plant. Inoculation may occur in many ways, but means of spreading diseases or inoculation can be classified into *soil-borne, wind-borne, seed-borne,* and *vector-transmitted* means. Much can be learned about the control of diseases by being aware of mechanisms by which various diseases are spread. Some diseases may be spread by more than one means.

Soil-Borne Diseases

Soil-borne diseases have to be introduced into the soil in some manner; the diseases do not appear in soil unless the soil is inoculated. Inoculation of soil may occur by wind, infected seeds, infected transplants, other infected plant parts, contaminated tools, water, or by other means that bring the soil in contact with the disease-causing organism. The organism survives and reproduces in the soil and infects crops that are grown in the infested soil. Crop residues in soil give excellent food on which the diseases can live. Soil-borne diseases may affect roots or underground organs or any part of the plant if the disease is spread from the soil to the foliage or transmitted within the plant. Some common diseases that are soil-borne are smut, bacterial wilt, fusarium wilt, verticillium wilt, club root, soft rot, stalk rot, root rot, damping-off, and potato scab.

Wind-Borne Diseases

Wind can carry diseases over great distances, principally by transporting spores of bacteria and fungi. Wind imports diseases that directly infect aerial portions of plants or that inoculate the soil. Some wind-borne diseases are various rusts, powdery mildew, downy mildew, apple scab, late and early blight of tomato and potato, and corn (*Helminthosporium*) blight.

Seed-Borne Diseases

Because of the possibilities of transmitting diseases by seeds, people are discouraged from saving their own seeds and encouraged to purchase certified disease-free seeds. Exceptions to saving of seeds may be in cases where the saved seeds are the only known or readily available source of a certain crop or where the saver has means of sterilizing seeds conveniently and safely. Also, saving seeds might be economical in some cases. Pod spots, bacterial blights, and anthracnose of peas or beans are spread by seeds. Several diseases of potato are spread by seed pieces (portions of tubers to propagate the plants). A few viral diseases may be spread by seeds to inoculate the soil in which vectors in turn lead to the infection.

Vector-Transmitted Diseases

Insects are vectors of several diseases and especially of the viral diseases. Many viruses are carried from one plant to another on the mouth parts of insects. Diseases can be spread by people who walk through gardens or fields and brush against or touch infected and uninfected plants. Pruning can spread viruses and disease-causing organisms. Soil-borne nematodes and fungi are also vectors for spreading of viruses, bacteria, and spores of bacteria and fungi. Mycoplasmic diseases can be transmitted by insects. Placing seeding in one's hands while planting seeds may transmit diseases to the seeds and subsequently to the crop.

INFECTION WITH DISEASES

Infection is the successful penetration and establishment of a disease-causing factor in a host plant. For infection to occur, the inoculum, as described above under "Spreading of Diseases," and a susceptible plant must be present. Susceptibility of a plant to infection means that the proper species in the right stage of development is present. Susceptibility of plants varies with environmental conditions. Plants weakened by poor nutrition, poor lighting, hot or cold temperatures, or by wet or dry soils may be rendered susceptible, thereby overriding resistance that the plants may have. Environmental conditions also must favor propagation of the disease for infection to occur. Warm, moist conditions in the soil or on the foliage can facilitate infection. If environmental conditions are not right for infection, the susceptible plant can be in contact with inoculum all season long without infection.

If the environment is favorable, infection may occur almost as soon as the plant comes in contact with the inoculum. Generally, bacteria, virus, and mycoplasma penetrate plants through wounds or natural openings in plant organs. Transplanting, pruning, mechanical damage from equipment, rough handling, bruising, puncturing, wind and rain, and insect or animal feeding give wounds through which diseases can enter plants. Natural openings include pores (stomata) in leaf surfaces, openings (hydathodes) at leaf margins, leaf scars from excised leaves, and pores (lenticels) in stems and fruits. Flowers have several openings through which inoculum may enter. Fungi, nematodes, and parasitic plants have the ability to enter plants directly through unopened surfaces or through natural or wound openings. Infection may occur on all of a plant, one side of a plant, or only on certain plant parts. The disease may be mostly external to the plant with structures entering the plant to gain nutrition for the disease. Other diseases may live within cells or between cells. With other diseases, the hyphae of the organism may be in the cells, and the fruiting bodies are outside the plant.

The time between successful infection and the appearance of symptoms is called *incubation*. The incubation period can be days as with bacteria or fungi or years as with certain viruses in woody plants. By the time that symptoms of diseases appear, much damage to the plants may have occurred. Diseases may be killed, but lesions resulting from the damage from the disease may persist.

CONTROL OF DISEASES

Control of diseases involves many procedures, grouped into activities that involve selecting disease-resistant plants, using cultural practices to increase or maintain plant resistance, keeping the causal organisms away from plants, and using pesticides.

SELECTING DISEASE-RESISTANT PLANTS

Use of disease-resistant plants should be given primary consideration in the selection of species and cultivated varieties of crops to grow. Some plants have structural features that impart resistance to some diseases. Thick waxy cuticles over leaves impede penetration of pathogens, particularly fungi, into the plant. A limited number and small sizes of openings (stoma, lenticils) offer some resistance to entry of bacteria into plants. Hairs on leaves limit wetting of leaves and deter infection. Some plants have disease resistance that is imparted by the biochemistry of the plant. Cells of some resistant plants naturally may release toxins that kill the pathogen. Some plants contain within their tissues certain chemicals that kill, slow the growth, or stop the action of diseases that penetrate into the tissues. Others may have a reaction of releasing disease-inactivating chemicals in response to wounding. Browning reactions with wounding are examples of this response. Chemicals in the plant may have an effect directly on the organism, whereas others may hinder the action of enzymes or toxins that are secreted by the diseases. Some plants have the capacity to metabolize toxins that are released by diseases and to prevent or limit the spreading of the disease in the plant. Resistance to viruses may be connected to the inability of the virus to be transmitted from cell to cell.

Crop scientists through breeding and selection of plants have developed many disease-resistant cultivars of crops. All growers should become familiar with the disease-resistant characteristics, if any, of crops and cultivars and, if possible or practical, make selections on the basis of these characteristics. In general, modern hybrids have resistance to many common parasitic and nonparasitic diseases or disorders that hinder open-pollinated varieties. Growers can obtain information about the disease-resistant characteristics of cultivars from plant names and from their descriptions in seed catalogs. For example, VF in a name indicates that a tomato cultivar is resistant to verticillium and fusarium wilts; VFN indicates resistance to these wilts and to nematodes. Descriptions of new hybrids may note resistance to different races of fusarium wilt (F_1, F_2), mosaic virus, cracking, tipburn, or other diseases and disorders.

CULTURAL PRACTICES TO INCREASE OR MAINTAIN PLANT RESISTANCE

Selection and Treatment of Plants and Seeds

Growers should always start with disease-free materials. Growers should purchase certified seeds. These seeds have been produced under conditions that are governmentally verified to be free of diseases. Growers may want to consider treated seeds. Some seeds are given a hot-water treatment (e.g., 120°F for 15 to 30 min.) to kill inoculum that is on the seed surfaces. Growers who save their own seeds may want to follow this practice. Bulbs also may be given the hot-water treatment. Bulbs also can be dipped in diluted household bleach to kill surface-borne diseases. A dilution of 1 to 5 or 1 to 10 (1 part bleach to 5 or 10 parts water by volume) is appropriate for this treatment. For many crops, seeds in lots sold in quantities larger than those small retail packets are treated with a fungicide, such as captan or thiram. These seeds might not be considered organic, and growers would want to check with their organic-certifying organization concerning the permitting of fungicide-treated seeds in a certified production system. Some vendors specialize in marketing seeds without fungicides.

Growers should inspect plants to ascertain that no diseases are present. Plants with swollen or knotty roots should be suspected of bearing club root or nematode infections. Other diseases, such as mildew, may be evident on foliage of transplants. Plants with mosaic pattering or puckering of leaves might contain viral diseases. With virus-prone plants (for example, strawberries, raspberries), growers should obtain certified (indexed) virus-free plants. A lot of care should be taken when obtaining plants from people who are not growers of certified seeds. Lettuce seeds should be indexed as being virus free. The terminology MT0 (MT zero) on a packet of lettuce indicates mosaic tested zero, that is, no virus on the seeds. Planters should avoid handling seeds with their hands. Many diseases are transmitted from hands to inoculate the seeds. Avoiding handling of seeds also lessens the exposure of the planter to fungicides on treated seeds.

Selection of Sites

Fungal and bacterial spores germinate well in warm, moist conditions. Although growers can do little to regulate the temperature of their gardens and fields, they can select sites that are well drained, well aerated, and sunny. Low-lying, wet areas should be avoided. Use of raised beds in these wet areas can help to deter infection by providing a well-drained area for planting. Pruning helps air circulation into plant canopies. Plants should not be spaced so close together that one falls on another or otherwise cuts off air circulation and drying of foliage or that walking through the planting is impeded by the plant canopy.

Rotation of Crops

Changing of crops on a plot of land can help to reduce disease infestation. In general, infestations of diseases will increase in soil with extended culture of the same crop on the same plot for several years, as the disease inoculum increases. Rotation of susceptible and nonsusceptible crops allows for reduction in infestation by reducing the density of inoculum. Crop rotation removes the susceptible crop from the land and allows time for residues of the crop to decompose. Ultimately, the pathogens die of starvation or of old age. Weakened pathogens are hastened to their death, because they are food for other soil microorganisms. Crop rotation is one of the best methods of control of diseases in large plantings, especially in the cereal crops (corn, wheat, rye, oats, barley).

Soil-borne diseases are relatively immobile compared to wind-borne or vector-transmitted diseases. Rotations can be successful in small plantings such as gardens. However, a considerable physical separation of crops or perhaps elimination of the susceptible crop from the garden for a season or more may be necessary for the rotation to be successful. Also, growers must be sure that the crops in the rotation are sufficiently unrelated so that the disease is not perpetuated. Crop rotations have little value in control of diseases that infect multiple hosts (such as fusarium and verticillium wilts).

Omission of most susceptible crops from land for 2 or 3 years will allow for decomposition of crop refuse (roots, stems, leaves, seeds) that contains pathogens. Longer omissions of 4 to 6 years may be needed for control of some diseases of brassicas, potatoes, and cotton. A period of fallow also may be included in the rotation to ensure that little or no organic matter is added from any source and that no host is in the soil to perpetuate the disease. Furthermore, frequent disking of fallowed land dries the soil and helps to diminish pathogens.

Rotations are not always required to diminish levels of pathogens in soil. Long-term monoculture of 4 years or more of the same crop can be a form of biological control of diseases, since in some cases, the population of a pathogen decreases with continued monoculture. In these monocultures, the pathogen becomes established after a few years of the same crop and causes severe losses, but with continued monoculture, severity of the disease declines and often disappears. A specific antagonism of other microorganisms to the disease apparently develops with monoculture. The antagonistic organisms are in a higher population in soil of the monoculture than they are in soil in which crop rotation is practiced. Patterns of decline of diseases with monoculture have been reported widely with take-all disease of wheat and with potato scab.

Control of Weeds

Weeds surrounding a garden or field may be hosts and source of inoculum for a variety of plant diseases. Many cultivated plants have wild relatives that as weeds in a field may serve as hosts to perpetuate diseases in a rotation. Club root of crucifer can be hosted on wild mustard, peppergrass, and shepherd's purse. Jimson weed and nightshade are hosts for diseases of solanaceous crops, such as tomato, potato, pepper, and eggplant. Common scab is carried on the amaranthus pigweed. Wild cherries have fungal diseases that may infect cultivated orchard crops. Pokeweed often is infected with viruses.

Some diseases require two, or alternate, hosts, spending part of their life cycle on one host and part on another. Alternate hosts for cedar-apple rust are cedar trees and apple trees; for white pine rust, the alternate hose is the black currant or gooseberry; and for wheat rust, it is the common barberry. Control of these diseases can be effected by eradication of the alternate host.

Control of Insects

Infestations of crops by insects can increase the susceptibility of plants to diseases in several ways, for example, by providing pathways for infection to occur, by weakening plants so that their resistance is diminished, or by spreading the diseases. Insects may puncture fruits or other plant organs and give pathways by which diseases can enter plants. Shoots of plants that have been weakened by insect attack often develop weakened root systems that have increased susceptibility to soil-borne diseases. This effect is noted particularly on trees that have been weakened by foliar damage by gypsy moths, leafrollers, or aphids or weakened by stem-boring insects. Injury to roots by insects or other forces likewise can weaken shoots and roots and increase susceptibility of both parts to diseases.

Insects are vectors for several plant diseases, particularly viruses and wilts. Aphids, leafhoppers, cucumber beetles, and other insects spread these diseases. Only a low population of insects is needed for this action. Far fewer insects are needed to spread diseases or to provide pathways for infection than are needed to weaken the plants or to provide direct damage.

Any method mentioned in Chapter 11, "Insect Control," might have value in lowering the spread of diseases by insects. To control the spreading of diseases by insects, however, the control method must be very effective. Growers may find use of row covers, reflective mulches, and sprays and dusts to be the most effective means for restricting spreading of diseases by insects. Proper timing of planting is important in avoiding spreading of diseases by insects. Early planting, or even fall planting of crops that will overwinter, can lessen dissemination of diseases. Plants infected while young remain sickly and small and provide favorable conditions for further feeding by insects, thereby leading to greater incidences of the diseases. However, early, well-established, large plants provide greater cover of soil, and some vectors, such as leafhoppers, are less likely to settle on a field covered by a crop than on a partially covered field. Also, the larger plants are likely to have lesser proportional infection than small plants, and feeding insects may be less apt to encounter infected foliage and to spread the disease to another plant.

Maintenance of Soil Fertility

Plants weakened by nutrient deficiency, acidic soils, poorly drained soils, or other conditions of poor soil fertility are more susceptible to infection by plant diseases than healthy plants are. Widespread nutrient deficiencies may go unnoticed and might not be significant until damage by a pathogen occurs. Fertilization with nitrogen helps plants to resist fusarium and bacterial wilts. Improved shoot and root growth as a result of improved nutrient availability enables plants to survive in the presence of inoculum. Fertilization of phosphorus-deficient plants has been shown to increase plant growth in the presence of soil-borne diseases, such as *Phytophthora*, even though the pathogen is not reduced by the fertilization. Enhanced mycorrhizal associations in roots after phosphorus fertilization have been suggested as providing plants with protection against infection. In some cases, greater shoot and growth following fertilization has been shown to lead to enhanced water removal from soil, thereby drying the soil and making it less favorable for pathogens.

However, an adequate water supply in the soil can lessen infection of crops. Irrigation is beneficial in deterrence of potato scab. Foliage of plants can be weakened by drought, leading the plants to be susceptible to diseases, such as powdery mildew.

Liming (pH 7.2) helps control club root of crucifer and bacterial wilt. It is difficult to maintain pH 7.2 in soils of humid temperate regions. Large or annual applications of lime will be needed for this practice to be effective. However, potato scab can be avoided in acid soils (pH 5.3), which limit the growth of the infecting organism.

Maintenance of soils at this pH is easy, unless the soils are calcareous with free deposits of limestone present. Additions of organic matter (e.g., peat moss), sulfur, ammoniacal fertilizers, or aluminum sulfate will bring about soil acidification.

Well-drained conditions help to avoid soft rots, bottom rot of lettuce, club root of crucifer, damping-off, and other soil-borne diseases. Circulation of air (oxygen) and drying of the soil help to deter growth of disease organisms.

Cultivation and Irrigation

Close cultivation that bruises plants provides means of entry of inoculum into plants. Proper spacing of plants and avoiding entry into the garden or fields when foliage is wet may lessen the damage imparted during cultivation. Deep cultivation in addition to pruning roots and limiting their capacity to absorb water provides means of entry of inoculum through broken and weakened roots. These weakened plants are highly vulnerable to infection from the soil or from above-ground inoculum. Drought-stressed, wilted plants are weakened and can be infected easily by windborne diseases, particularly powdery mildew. Irrigation can restore some disease resistance to these plants.

Care should be taken not to overwater plants. This caution is very important in the culture of container-grown plants and in the propagation of plants from seeds. Care should be taken not to splash soil onto leaves or fruits. Also, wetting of foliage may burden plants and cause them to fall in contact with soil. Some recommendations suggest avoiding watering of plants late in the day or at night so that the foliage of plants is not moist for prolonged periods during which infection can occur. This recommendation should be tempered by common sense. Plants noted late in the day or at night to be suffering from lack of water should be watered. In some cases, efficiency in use of irrigation water is achieved by watering at night when evaporation rates are low.

As much as is possible or practical, growers should avoid application of water from sprinkler systems or water from hose nozzles on plants, as wetting of foliage can increase infection of plants. Wetting of foliage can break weak stems or force upright plants to lodge. Wet foliage provides an environment suitable for disease infection. Growers should try to apply irrigation water directly to soil. In cases where direct application to the soil is impractical or not feasible, as in some dense plantings or in lawns, growers should employ practices that enable the foliage to dry soon after irrigation.

Use of Protective Fungicides

Fungicides help to protect plants by providing a protective barrier to prevent infection, by killing spores or germinated spores, or by killing the disease after infection. Early treatment with fungicides is critical for disease control and for recovery of the crops. The organic grower has a few protective fungicides for use in preventing or controlling disease infections. Agricultural limestone, slaked lime, and quicklime are fungicidal. The caustic action from the alkalinity of these products offers some protection against germinated spores and fungal bodies. Wood ashes, also being alkaline materials, can be effective in disease control. The proteins in milk make it a mild fungicide. Sulfur is an organic fungicide. Its effect is in inhibition of spore germination. Limes (or wood ashes) and sulfur may be mixed for the effectiveness of inhibition of germination and killing of germinated spores and infections. Copper hydroxide is approved for organic crop production. Some organizations permit use of copper salts (copper sulfate) or copper salt and lime mixtures. Copper-lime mixtures are called *Bordeaux mix*. Lime-sulfur and Bordeaux mixes are not made in exact proportions and can be made in several formulations for dusting, spraying, or sprinkling on plants. Either mix can have some phytotoxicity, particularly during hot weather and also may affect other nontarget organisms in the environment. The copper-containing fungicides may have restrictions in their use. These restrictions may include proper protection of the persons doing the application and procedures that minimize copper accumulation in soils.

Alcoholic beverages, mouthwashes, molasses, and hydrogen peroxide used alone or in combination have been suggested as fungicidal materials. Growers should test these materials for their efficacy in disease control and should ensure that these materials are not toxic to the plants being treated. The efficacy of these materials may be related to their surface sterilization of foliage, to enhancement of competitive microbial growth on foliage, or to other suppressive effects on the growth of disease organisms. Some of these materials, such as hydrogen peroxide, kill only by contact and have no residual or curative action.

Sanitation

Diseased plants should be removed and destroyed as soon as the disease appears. Pruning out the diseased portions may be satisfactory if removal of the entire plant is not practical. Removal of dropped leaves and fruits is helpful in breaking the life cycle of diseases. In general, it is unwise to compost diseased plants or parts of plants. The temperatures in compost piles become high enough in the inner zones of the piles to kill fungi and bacteria. Temperatures of about 140°F quickly kill fungal and bacterial diseases. Temperatures of 105°F will kill these diseases in several weeks time. Toxins generated during composting also may kill these diseases. However, not all of the diseased material may be exposed to the hostile environment of high temperatures and toxins, and the diseases may live in the outer parts of the piles. Tobacco mosaic virus (TMV) and some other viral diseases are not killed by composting. Temperatures of about 180°F are needed to kill viruses. However, except for TMV, viral diseases are not likely to be spread by compost because of the lack of a vector to bring about the infection. However, TMV does not require a vector for infection.

Crop residues should be turned into the soil after the harvest season. The organic matter will rot in the soil, thereby hindering the overwintering of the diseases. Fallen leaves from mature ornamental trees or shrubs and fruit trees or brambles should be removed. Covering of the fallen leaves with a layer of compost physically may inhibit spreading of diseases from the fallen leaves to the canopy of the trees or shrubs in the next season. Mixing of compost into the soil may introduce microorganisms that compete with disease organisms sufficiently to reduce populations of the disease-causing organisms to noninfectious levels as a form of *biological control*.

Tools should be kept clean. Soil clinging to spades, trowels, hoes, and other garden tools should be scraped off, and the tools should be washed with water after their use. Pruning tools should be cleaned between plants and, if infected portions are pruned, between portions of plants. Pots and seeding flats should be cleaned and washed with detergent before they are used. A dilute bleach solution (1:10 dilution of household bleach with water) also can be used in disinfecting tools, pots, and flats. This solution also can be used to dip bulbs in before they are planted from one part of the garden to another. Pruning knives can be disinfected by dipping in rubbing alcohol (isopropanol) between uses.

Biological Control

Any control in which a live antagonist of the disease is involved can be called *biological control.* Biological control rarely eliminates a pathogen from the environment but instead reduces the numbers of the pathogen or restricts its ability to produce diseases without an actual reduction in its population. Practices and interpretation of biological control differ widely among growers. Some growers consider the use of resistant varieties to be a form of biological control. Eradication of alternate hosts and destruction of infected plants also are forms of biological control. Crop rotation also is biological control. In the context of this book, however, biological control involves intentional introduction of the antagonist of a disease or involves the creation of conditions that give the antagonist an advantage over the disease.

Beneficial microorganisms in the environment help to control plant diseases by competition with the diseases, by production of antibiotics or lysing of metabolites that affect the growth of the diseases, or by infection or predation of the diseases. Some soils naturally have a complex population of microorganisms that interact to suppress plant diseases. A fertile soil is more likely to have this complex population than a poor soil. Good practices of land and soil management, such as drainage, intensive cover cropping (green manuring), liming, and fertilizing, help to build up this diverse population of microorganisms. The suppressing action of beneficial organisms can be overcome by waterlogging, adding large amounts of nitrogen-rich organic matter, or by input of large amounts of disease inoculum. A waiting period of a few days or weeks between incorporation of nitrogen-rich organic matter, such as farm manures or organic fertilizers, and planting or seeding of crops should be provided to allow for dissipation of diseases stimulated by the addition of organic matter.

Moderate amendments of organic matter to soil help promote biological control. The total intensified microbial action induced by the organic matter seems to be involved in biological control of diseases. Also, organic matter additions to soil might increase the dormancy propagules (e.g., spores) of diseases and increase their digestion by other soil organisms. These actions are not highly specific to kinds of organic matter added, to diseases, or to antagonistic organisms.

Generally, the greater and more diverse the population of microorganisms supported by amendments with organic matter, the greater the probability of a disease-suppressing effect from the amendments. Plowing under green manures or additions of farm manures have been shown to lessen infections with potato scab, root rots, and take-all disease. Green grass clippings have been shown to promote control of potato scab. The kind of green manure (legumes, grasses, or broadleaf nonlegumes) is usually not a large factor in the disease-suppressing activity of adding organic matter to soil. However, legumes may be the preferred green manure for disease control.

Leguminous residues because of their rich nitrogen content, as well as carbon, stimulate intense biological activity. Ammonia and other products of decomposition of legumes may inhibit or not support continued growth of disease organisms.

As mentioned previously, growers should take caution against planting too soon after a green manure or other uncomposted organic matter, including organic fertilizers, has been incorporated. The microbiological activity induced by the green manure and other fresh organic matter and the metabolic products of this activity can kill, weaken, or make crops susceptible to diseases.

Green manures can be grown as a cover crop over winter and incorporated each year before the main crop is planted. In some climates, legumes may not be suitable for this activity, thereby limiting the cover crop to rye or other cereals. The nonleguminous cover crop should be turned under while green, for strawy mature crops will immobilize nitrogen and lessen microbial activity, particularly that of the microorganisms that are antagonistic to diseases. Contrary to this response, mature straw of cereal grains incorporated shallowly into soil has been demonstrated to starve pathogens of nitrogen. Roots of the host plant penetrate the soil to depths below the incorporated straw and are not affected by the shallow immobilization of nitrogen.

With incorporation of green manures in the soil annually, monocultures of cash crops are possible. The added organic matter stimulates antagonistic responses sufficiently to control pathogens to levels at which crop production is not hindered by diseases. Growers should understand that a long term may be needed for the balance in favor of crop production to be achieved and that for a few years diseases may not be well controlled. Also, growers should recognize that imbalances may occur as the result of larger than normal incorporations of organic matter, with wet or dry seasons, or with other uncontrolled environmental factors. Growers starting out with use of organic matter amendments to control disease should consider that moderately nitrogen-rich, rapidly decomposing matter stimulates the greatest number of soil microorganisms and gives the best control, whereas slowly decomposing materials give lesser stimulation of microorganisms and lesser control. This relationship does not hold in all circumstances, because in very infertile soil low in nitrogen, other nutrients, and organic matter, additions of rich organic matter initially can stimulate growth of pathogens more than the growth of antagonistic organisms. Also, following applications of organic matter, particularly nitrogen-rich matter, the growth and demands of beneficial organisms may be so intensive that these organisms become temporary plant pathogens.

Incorporation of compost encourages growth of beneficial microorganisms in the soil. Compost contains and promotes growth of large populations of organisms (e.g., *Trichoderma* and *Gliocladium* fungi) that suppress fungal and bacterial diseases and nematodes.

Use of trap and disease-inhibitory plants has been tried in nematode control. The practice of using trap plants involves growing a highly susceptible crop to catch nematodes. The infested trap plant is removed before egg masses are deposited. Timing and the hazards of increasing the population of nematodes are problems with this practice. Some work has demonstrated that the use of highly resistant crops results in better nematode control than the use of susceptible crops, because of the lack of a good host in the resistant plant and subsequent starvation of the disease. Inhibitory crops might reduce the population of nematodes. Marigolds (French, Mexican, or African *Tagetes* spp.) are resistant to root-knot nematodes. Green manure crops of marigolds have been reported to cause reductions in nematode populations in soil, and fewer nematodes were detected in infested soil near marigold roots than near other kinds of plants. Marigolds produce a highly nematode-suppressing chemical that can be released even without decay of roots. However, 3 or 4 months may be required for sufficient root growth for nematode-controlling effects by marigolds to be evident. The effects from marigolds might be exerted on other diseases, such as the wilts *Fusarium*, *Verticillium*, and *Helminthosporium*, but this control is not as well documented as that on nematodes. Other plants that have been shown to have hindering effects on nematodes are asparagus (stubby-root nematode) and rape or mustard (golden nematode).

Most of the preceding discussion on biological control has centered on the stimulation of microbial activity by making environmental conditions favorable for their growth. It is also possible to inoculate seeds with disease-suppressing organisms, much in the same way that seeds of legumes are inoculated with *Rhizobium* prior to sowing. Inoculation of cereal (corn, barley, oats, wheat), marigold, and carrot seeds with bacteria (*Bacillus subtilis*) or fungi (*Streptomyces* sp., *Penicillium* sp., and others) may offer protection against several soil-borne rots, wilts, and blights and enhanced yields and flowering responses. Introduction of the suppressing organisms directly into soil and use of soil enriched with these organisms have been applied to disease control with some success. This practice has been used in the control of damping-off disease.

Biological control of soil-borne pathogens has been emphasized here. This emphasis is justified because of the large number of soil-borne diseases that infect below-ground and aerial parts of plants. Some of the wind-borne diseases (fire blights, rusts, mildews) and vector-transmitted diseases (aphid-borne and leafhopper-borne viruses, Dutch elm disease, chestnut blight) have no connection with the soil. However, evidence suggests that a large and diverse microbiological population exists on the aerial parts of plants and that attacks by pathogens may be suppressed by other microorganisms on plant surfaces. Perhaps the use of sprays of molasses on leaf surfaces enhances diversity of microorganisms and increases competition with the disease-causing microorganisms.

Despite their potential, development of biological practices for control of diseases of plant shoots has been limited. Perhaps a limiting factor in this development is the fact that application of biological controls is in direct competition with use of various foliar sprays and dusts, which are deemed to be more effective.

Pathogens entering plants through wounds may be suppressed by saprophytic microorganisms residing in the wounds. Application of disease-suppressing fungi to fresh wounds of fruit and forest trees has been practiced successfully in control of diseases. Applications of antagonists controlled *Botrytis* spp. fungal disease on lettuce and tomato. Mounding of soil around, but not covering, aerial portions of lettuce also reduces botrytis infections. Research has shown that it is important that the antagonist be present when the pathogen strikes the aerial portions. If the pathogen strikes first, it occupies the tissues and inhibits the effects of antagonists.

Solarization

Solarization refers to the practice of covering soil with a layer of clear polyethylene film, which traps solar energy and raises soil temperatures high enough to kill or inhibit plant diseases, insects, and weeds. Typically 3 or 4 weeks of solarization are applied for disease and insect suppression, but up to 6 weeks may be needed for weed suppression. Time is needed for soil temperatures to be raised sufficiently and deep enough to be effective. Temperatures in the solarized zone should rise to about 110°F and remain there for several weeks. The effective depth of warming should be 4 to 8 inches. Since soil warms from the surface downward, soil at the 4-inch depth will be warmed for about twice as long as that at the 8-inch depth. So, after solarization, the soil should not be cultivated more than 3 to 5 inches deep to avoid the possibility of bringing up pests from the unaffected depth zone.

The area to be solarized should be wetted thoroughly before covering with the plastic. After solarization is finished, the plastic may be removed or left in place and covered with latex paint, whitewash, or an organic mulch to exclude light and to impart weed control. Mixing manure or compost in the soil before covering for solarization improves control of weeds that might be transmitted with these materials. Placing a layer of dark-colored compost on the soil surface before the plastic is applied increases the warming temperatures of the soil by several degrees. Solarization works best in warm summer months and in warm climates. Warming of solarized soil in the spring in northerly climates probably will be insufficient to kill pests.

Storage

Only sound, well-matured, externally dry, disease-free fruits and vegetables should be stored. Produce that is punctured should not be stored. The stem portion of tomatoes should be removed at harvest so that one fruit does not puncture another, although in some practices the stem is left on the fruit as a factor in marketing, as some people believe that the remaining stem is an indication of freshness. Some curing of produce after harvest helps to improve storage life of the produce. Onions cured in the sun for a few days store better than uncured onions. The storage area should be cool, well aerated, and ventilated. Whitewashing of the interior will help to disinfect the storage area.

13 Companion Planting

Companion planting is the *interplanting* of two or more crops that benefit from their association. The beneficial effects may be exerted in several ways. Reported benefits of companion planting are control of pests, improvement of soil conditions, improvement of product and yield, and thinning of crops. Some of the benefits are well documented, whereas others are untested and may be only testimonials.

Some people believe that companion planting is an exact process. The *biodynamic gardener's* approach to companion planting considers which plants grow well with one another and which do not grow well with one another. Extensive recommendations have been developed for interplanting based on the putative effects of plants on one another. In some views, this approach is an extreme one in that the precision and details of recommendations go beyond the expected effects that plants normally have on one another. Nonetheless, a considerable amount of planning should be done in selecting plants for interplanting if companion planting is to be practiced in crop production.

In interplanting, plants compete for the same space and for the light, nutrients, and water provided in the space. Consideration must be given to the relative vigor and size of plants for interplanting. Tall plants may shade shorter plants, thereby aiding or suppressing the shorter plants. Established plants give winter protection to young plants that are not yet established or acclimated to cold temperatures. Some plants have vigorous roots systems that penetrate deeply into the soil or that expand extensively laterally. These luxuriant root systems can help in loosening soil for the growth of other plants in the interplanting. However, the roots of one plant may exhaust nutrients and water from the soil and deprive other plants of nutrients and water.

Grasses and legumes are grown frequently in interplanting. The benefits to the grower are usually achieved in a better product (see the section titled "Improvement of Product") rather than in mutually beneficial effects of the plants on one another. Care must be taken to ensure that growing conditions for the combined planting are optimum rather than having conditions that favor an individual crop.

Some of the benefits of interplanting of grasses and legumes have been ascribed to the fact that legumes fix nitrogen from the air. The mutual benefit from nitrogen fixation is small. Some nitrogen may be excreted by the living roots of legumes, but usually not enough nitrogen is excreted so that grasses or other nonlegumes living in association with legumes receive benefit. The benefits from the fixed nitrogen are realized only after the leguminous plant dies and its nitrogen is released by mineralization. Nonlegumes interplanted with living legumes can be nitrogen deficient, because the living legumes are contributing no available nitrogen to the soil and are competing with the nonlegumes for soil nitrogen. Three-fifths to two-thirds of the nitrogen in a legume is in the shoots rather than in the roots as many people perceive.

The relative densities of planting of legumes and nonlegumes must be balanced, possibly with a higher density of the nonlegume, to avoid giving a competitive advantage to one species or the other in the absence of fertilization. However, the response of a mixture of grasses and legumes to nitrogen fertilization is production of luxuriant growth of the grasses, growth which outcompetes that of the legumes thereby leading to a reduction of the proportion of legumes in the mixture. Nitrogen fertilization of pastures and hay land consisting of mixtures of grasses and legumes can lead to domination of the grasses to the point of exclusion of the legumes.

Some plants have effects on others by means other than direct competition for light, water, nutrients, and space. The noncompetitive, suppressive effect of one plant on others in an association has been studied considerably and is called *allelopathy*. In choosing crops for interplantings, growers also must be aware of allelopathic effects that crops may have on one another. The effects may be exerted by compounds exuded from the roots of a donor plant. Also, compounds may be contained in plant parts other than roots and may be delivered to the ground by dripping, by rainfall, by leaf drop, or by tillage of the aerial portions into the ground. The effect of black walnut trees on underplantings is a well-known example of allelopathy. Black walnut trees suppress the growth of ornamental shrubs, blueberries, tomatoes, and other crops that are under the canopy of the trees. A compound called *juglone* is transferred by dripping from the leaves of the walnut tree onto the foliage of plants under the trees. The allelopathic effect of trees on crops should not be confused with the competitive effect that the trees have for water.

Many common row crops have allelopathic effects on plants. Sunflower roots appear to exude compounds that inhibit growth of other plants, including weeds. In addition, residues of vegetative portions and seeds of sunflower are allelopathic. Allelopathic effects are associated with corn residues, straws of small grains, soybean residues, among other crops, and weed residues. Allelopathic effects may be used in weed control, provided that the allelopathy is not exerted on the main crop. Timing of planting can be a factor in benefiting from allelopathy among plants.

BENEFITS OF COMPANION PLANTING

INSECT CONTROL

Some plants are reported to repel insects and similar pests. Much of this alleged activity is untested, but some examples include reports of catnip deterring flea beetles, mint driving away cabbage looper moths, garlic repelling Japanese beetles, and marigolds killing soil-borne nematodes. Evidence for marigold killing of nematodes is the best documented effect among this group. Some plants attract insects and serve as trap crops on which the pests can be killed or removed, leaving the main crop intact and perhaps uninfested. Solanaceous (members of the family Solanaceae) weeds or crops may be used to trap potato beetles. Corn is used to attract earworms, which are the identical pest as tomato fruit worm and tomato bud worm. Turnip, radish, and mustard have been used to trap harlequin bug from cabbage, lettuce, corn, eggplant, squash, and potatoes. Some plants are said to attract beneficial insects, spiders, and mites into a planting and to help impart some biological control of pests. Asters are reported to attract beneficial spiders. Fennel is reported to attract beneficial wasps, and petunias are reported to attract lacewing.

Assessments of the value of companion planting in insect control suggest that most plants have very little effect in controlling insects on other plants. The success of some interplantings appears to be due to the fact that some crops have the potential to yield fairly well in the absence of insect control of any kind, whereas the failure of some interplantings is due to the fact that some crops cannot be grown successfully without use of a method of insect control.

Improvement of Soil Conditions

Some crops such as alfalfa are deep rooted. Their roots can penetrate considerably into the soil and help to break up pans, to crumble the soil, and to provide channels for improved drainage. Growers may want to avoid growing deep-rooted crops in dry land because of the probability of depletion of water in the ground by plant transpiration and the difficulty in recharging the soil with water. Depletion of soil water would be detrimental to any interplanted crops or to crops in rotation with the deep-rooted species. However, plants with fibrous root systems can help with loosening and crumbling of soils at shallow depths. However, the benefits of crops on improving soil structure may not be obtained in interplantings as well as they are obtained in crop rotations in which a green manure is grown and turned into the ground.

Improvement of Product

Forages of legume-grass mixtures are often of better quality for feeding of livestock than forages of the crops grown singularly. Growing and harvesting forage from an interplanting of legumes and grasses leads to a balance in protein, carbohydrate, and fiber in the product. The legumes enrich the mixed forage with protein. The grasses enhance the fiber and carbohydrates in the forage. Legume forage alone is protein-rich and may upset the diet of livestock feeding on purely leguminous forage. Grass forage is low in protein, and livestock feeding on strictly grass forage or grain usually require a protein supplement. The legume provides this supplement.

Thinning of Crops

Gardeners may want to consider interplanting of two crops, one of which is harvested before the other, leaving the remaining crop properly spaced for high yields. It is important in these interplantings that the crops have quite different rates of maturation to a harvestable product and that the competition between the crops is not intense. Carrots and table beets or radish can be interplanted with the beets or radish being harvested before the carrots. With interplantings of carrots and leeks, the leeks occupy space so that the carrots are not too thick. In this case, the carrots are the first crop to be harvested.

Weed Control

The possibility of weed control by companion plantings is highly speculative. Often one of the crops acts as a weed in relation to the other by competing for space, light, water, and nutrients. Often a grower may plant a crop in a dense planting of another species and then use some method to suppress the growth of the second species. For example, gardens can be planted in sod in which only enough area is tilled for the main crop, and the sod between rows is mowed and controlled to provide pathways and to improve the appearance of the garden. Weed control is best achieved by use of an allelopathic crop in crop rotation or by importing residues of that crop onto the field or garden. Residues of oats and rye have been demonstrated to give 50% to 70% control of barnyardgrass, purslane, and crabgrass. Similar controls have been shown for corn residues on growth of lambsquarters and redroot pigweed. Residues of marigold have been reported to control purslane by 80%. Although none of these levels of control are sufficient for weed suppression (about 95% control is needed), use of weed-suppressing residues may make the job of weed control easier, particularly if the grower is using cultivation as a means of weed control.

Plans for Companion Planting

Interplanting of crops can be accomplished in many designs. A common method of interplanting is to alternate plants in the same row (Figure 13.1a). This pattern puts each plant of one species next to another plant of a different species and gives a high probability of interaction between species. Also, several plants may be planted in a

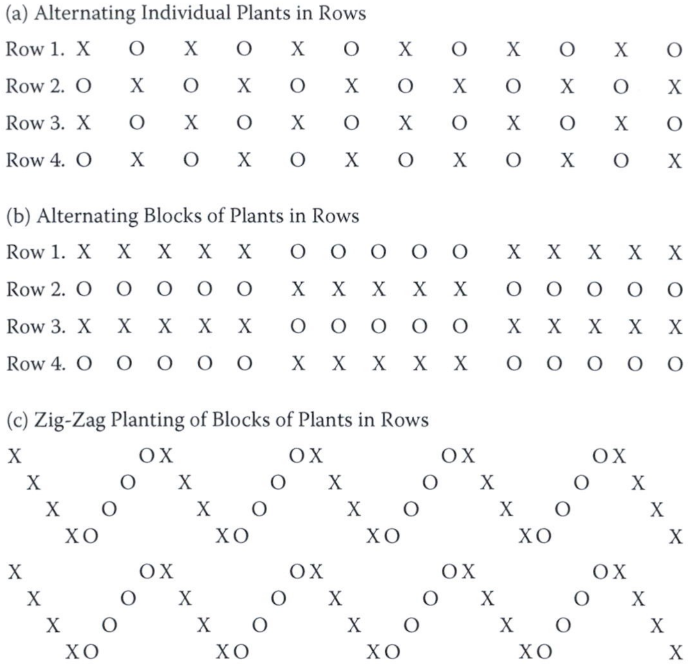

(a) Alternating Individual Plants in Rows

Row 1.	X	O	X	O	X	O	X	O	X	O	X	O
Row 2.	O	X	O	X	O	X	O	X	O	X	O	X
Row 3.	X	O	X	O	X	O	X	O	X	O	X	O
Row 4.	O	X	O	X	O	X	O	X	O	X	O	X

(b) Alternating Blocks of Plants in Rows

Row 1.	X	X	X	X	X	O	O	O	O	O	X	X	X	X	X
Row 2.	O	O	O	O	O	X	X	X	X	X	O	O	O	O	O
Row 3.	X	X	X	X	X	O	O	O	O	O	X	X	X	X	X
Row 4.	O	O	O	O	O	X	X	X	X	X	O	O	O	O	O

(c) Zig-Zag Planting of Blocks of Plants in Rows

```
X        OX        OX        OX        OX
 X      O  X      O  X      O  X      O  X
  X  O      X  O      X  O      X  O      X
   XO        XO        XO        XO        X
X        OX        OX        OX        OX
 X      O  X      O  X      O  X      O  X
  X  O      X  O      X  O      X  O      X
   XO        XO        XO        XO        X
```

FIGURE 13.1 Diagram of planting plans for alternating rows of crops in rows. (a) Alternating individual plants (Xs and Os) in rows, (b) alternating groups of plants in rows, and (c) alternating groups of plants in zig-zagged row.

group in a row (Figure 13.1b). This planting may have advantages of putting several of the same kind of plants together to aid in pollination and in harvest. The rows may be zig-zagged to bring the different species in closer proximity with one another than would occur in a strictly linear planting, especially if only one row is planted (Figure 13.1c). If two or more rows are planted, placement of crops could be alternated among the rows to ensure diversity in planting (Figure 13.1a,b). Plants may be in sections or blocks containing several rows of one species that are alternated with sections or blocks of another species (Figure 13.2a). This planting pattern gives groupings of species to facilitate pollination and harvest. A border of plants around a garden or plot is a form of companion planting (Figure 13.2b). Borders may be used to attract beneficial insects into the area or may be a place for growth of trap crops.

PITFALLS OF COMPANION PLANTING

Many of the statements of success in use of companion planting come from unsubstantiated testimonials. Growers should beware of claims such as "Carrots dislike dill," "Scarlet sage repels rodents," and "Radishes are not happy if hyssop is close by." Sometimes the beneficial factor is not recognized or is misjudged. Many of these misconceptions as to the beneficial factor are associated with weeds. Many growers expecting a beneficial effect from weeds intentionally allow weeds to grow with their crops. Flowering weeds may attract pollinating insects into areas and help in setting of fruits by cross-pollinated crops. Some people suspect, however, that in some cases flowering weeds may attract bees away from crops and restrict pollination. One should always be concerned with the competition of weeds with crops with

(a) Alternating Blocks of Plants

```
XXXXX OOOOO XXXXX OOOOO XXXXX OOOOO
XXXXX OOOOO XXXXX OOOOO XXXXX OOOOO
XXXXX OOOOO XXXXX OOOOO XXXXX OOOOO
XXXXX OOOOO XXXXX OOOOO XXXXX OOOOO
XXXXX OOOOO XXXXX OOOOO XXXXX OOOOO
XXXXX OOOOO XXXXX OOOOO XXXXX OOOOO
XXXXX OOOOO XXXXX OOOOO XXXXX OOOOO
```

(b) Border of Plants around a Block

```
OOOOOOOOOOOOOOOOOOOOOOOOOOOOOOOOOOOOOOOOOO
OXXXXXXXXXXXXXXXXXXXXXXXXXXXXXXXXXXXXXXXXXXXXXXXO
OXXXXXXXXXXXXXXXXXXXXXXXXXXXXXXXXXXXXXXXXXXXXXXXO
OXXXXXXXXXXXXXXXXXXXXXXXXXXXXXXXXXXXXXXXXXXXXXXXO
OXXXXXXXXXXXXXXXXXXXXXXXXXXXXXXXXXXXXXXXXXXXXXXXO
OXXXXXXXXXXXXXXXXXXXXXXXXXXXXXXXXXXXXXXXXXXXXXXXO
OXXXXXXXXXXXXXXXXXXXXXXXXXXXXXXXXXXXXXXXXXXXXXXXO
OXXXXXXXXXXXXXXXXXXXXXXXXXXXXXXXXXXXXXXXXXXXXXXXO
OXXXXXXXXXXXXXXXXXXXXXXXXXXXXXXXXXXXXXXXXXXXXXXXO
OOOOOOOOOOOOOOOOOOOOOOOOOOOOOOOOOOOOOOOOOO
```

FIGURE 13.2 Diagram of planting plan for (a) alternating blocks of crops and (b) placement of a border row around a block of a crop.

respect to availability of water, nutrients, and light. Weeds are places where insect pests may harbor. Weeds are also good sources of disease inoculum. Claims that best harvests of pumpkins, or other crops, come from areas in which jimson weeds grow have associated a benefit from the presence of the jimson weeds. Jimson weeds grow best in very nutrient-rich soil, such as old barnyards. The presence of jimson weed is an indication of the high fertility of the land, and crop yields are high in response to this fertility. Yields would be even higher if the weeds were removed. Good yields of high-quality watermelons, or other crops, have been associated with weedy patches of land. Again, the weeds are growing well in response to soil fertility, which might include factors such as moisture as well as nutrients. One might concede with watermelons that some shading by the weeds allowed the melons to be cool and to taste better on fresh harvest than ones in full sun.

14 Storage of Produce

In this chapter, storage of fruits and vegetables without refrigeration is presented. Storage without refrigeration requires a proper facility, such as a basement, a cellar, an outbuilding, or a pit. The kind of facility used depends on the climate in the area, but generally, most unrefrigerated facilities are not practical unless the outdoor winter temperatures average 32°F or lower. If the outside temperature is not cool enough, temperatures in the storage area will be too warm to prevent spoilage of produce other than dried materials.

STORAGE FACILITIES

BASEMENTS

Without modification, an ordinary basement under a house with central heating from a furnace placed in the basement can be used for short-term (a few weeks) storage of potatoes, sweet potatoes, and onions and for ripening of tomatoes. Electric heat in a house usually ensures a cold basement, but furnaces in basements will emit too much heat for long-term (over winter) storage of fruits and vegetables. Basements with furnaces will need to be partitioned to provide a cool storage area. Preferably, the partitioned area should be in the coolest portion, which is usually the northeast corner or on the north or east side of the basement. No heating pipes or ducts should run through the room. If these units are present, they should be insulated. The storage room should have at least one window for cooling and venting the room. Odors can be a problem if they are allowed to seep into the living areas of the house, adding another essential requirement for good ventilation in a basement storage area. Because of odors that may permeate into living quarters, turnips, rutabagas, cabbages, and possibly onions should not be stored in basements.

If the storage room is divided in half for storage of fruits and vegetables separately, provide one window for each division. The windows should be shaded to keep light from the storage room. The room should be fitted with shelves and slatted floor pallets to keep produce off the floor and to allow for air circulation. Sawdust may be put on the floor and under the pallets and dampened to raise humidity. Fruits and vegetables should be stored in small wooden crates or boxes that are placed on the shelves or pallets rather than in large bins.

CELLARS

Cellars are unheated structures under houses or underground or partly underground structures away from houses. Cellars under houses might have separate entrances from the house, thus distinguishing them from basements. The door to the entrance gives a means for ventilating and regulating the temperature in the cellar. Cellars for storage under a house should be managed the same as the partitioned room of a basement.

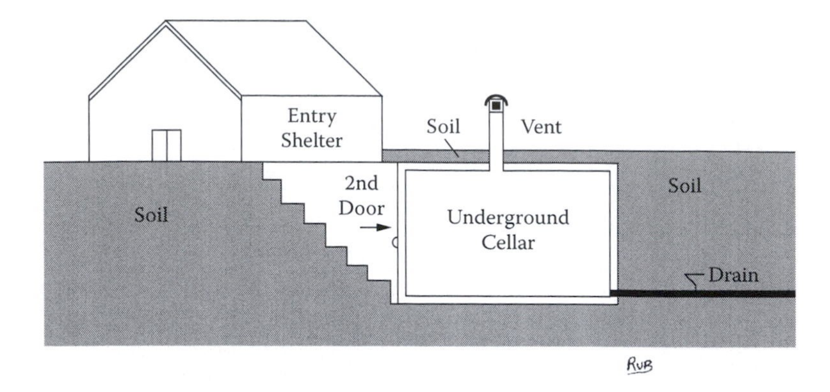

FIGURE 14.1 Diagram of an outdoor, fully underground cellar with above-ground entry shelter.

Outdoor cellars can be expensive to construct; hence, only limited space is usually available in them. They provide excellent facilities for storage of produce in cold areas and give relatively uniform temperatures year round. For the fully underground cellar (Figure 14.1), the walls and roof must be strong enough to support the soil that covers them. Reinforced concrete provides the strongest structure, although concrete blocks and stones for walls have been used in combination with concrete roofs.

In many parts of the country, these structures were built as storm cellars, particularly at schools, as shelters from tornadoes.

Outdoor cellars are usually windowless, unless some portion extends above ground or is exposed on one side, and ventilation is provided by flues and by opening of doors. An above-ground structure provides the entry with an outside door. At the floor level of the cellar, another door is provided. The second door is important to provide insulation against cold air that enters through the entry way of the above ground structure. The partly underground cellar is constructed with masonry walls that are surrounded on three sides by soil (Figure 14.2). The entry way is directly from the outdoors.

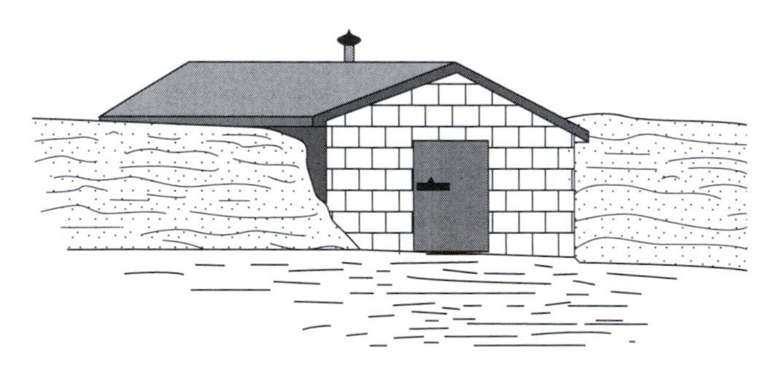

FIGURE 14.2 Diagram of partially underground cellar with back and side walls covered with soil. With concrete masonry construction of the roof, the cellar can be covered with soil, with only the front being exposed fully to the outdoors.

The fully underground cellar, outdoors or under a house, can be used for storage of canned fruits and vegetables. Freezing is not unavoidable in cellars, and growers should be aware of the possibility of loss of canned goods on cold nights, particularly if the doors are not closed and well sealed.

Since cellars are cooler in summer than the above ground areas, they can be used for storage of produce harvested during warm weather and for which cool storage is beneficial (for example, onion bulbs).

Outbuildings

Storage of fruits and vegetables in outbuildings (Figure 14.3) is practical only in climates with consistently cold temperatures averaging near or below freezing during the period of storage. Unlike the basements or cellars, with outbuildings some supplemental heating may be required on cold nights. Outbuildings can be built of many different materials, including masonry, wood, and metal, but they should be insulated. Hollow walls have little insulating value and should be filled with vermiculite, rock wool, or some other dry, granular material. A vapor barrier should be included between the inside walls and the insulation. The moisture barrier can be aluminum paint or foil, plastic, or tar paper. The outside walls should be made tight with building paper covered with wood or other kinds of siding. The building should be ventilated with intake and exhaust vents with or without fans.

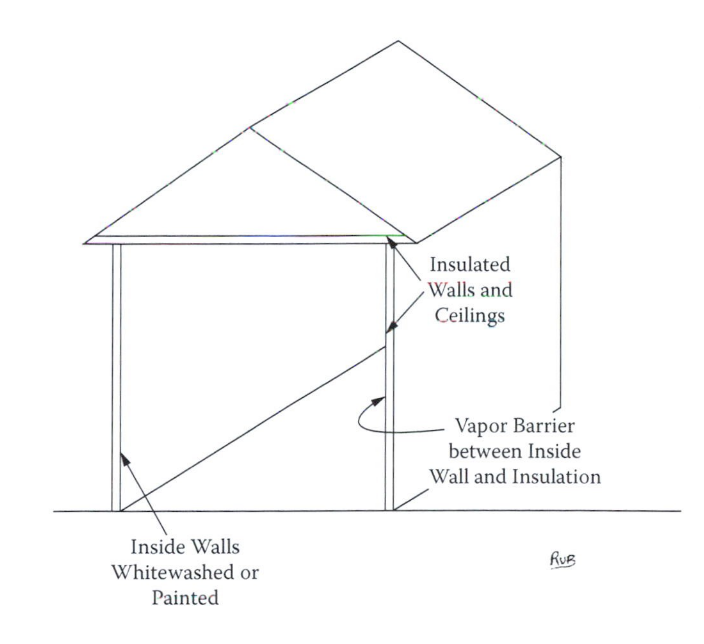

FIGURE 14.3 Structural design of a fully above-ground outbuilding for storage.

Pits

Pits are outdoor structures constructed largely of soil. They are not deep holes in the ground as the name might imply. Pits usually are built on top of the ground or start in a hole about 6 to 8 inches deep. They may be cone-shaped or long-shaped (Figure 14.4). Cone pits are used for storage of small amounts of produce to be removed all at once, whereas long pits can store large amounts of produce that can be removed over a period of time. Root or tuber crops (potatoes, carrots, beets, turnips, parsnips), cabbage, apples, and pears are among the produce that can be stored in pits. Pits start with a layer of leaves, straw, sawdust, or woodchips on top of the ground or at the base of the 6-to-8-inch deep hole. The produce to be stored is placed on top of this bedding. The produce should be covered with more straw or other bedding. Except for a portion of the bedding that extends through the cover, the entire mass should be covered with about 4 inches of soil, which is firmed to help make the pit waterproof. Extension of the bedding through the soil allows for ventilation of the pile. The top of the pile where the bedding protrudes should be covered with a sheet of metal or a board, which is held down with a stone or some other weight. A loosely placed and anchored sheet of plastic also will make a top covering. A trench should be dug around the pit to provide drainage away from the storage pile.

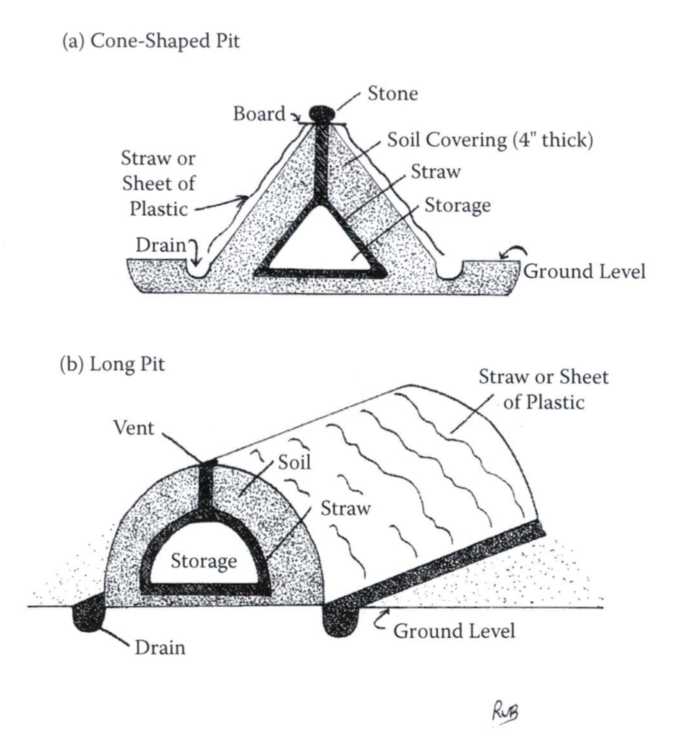

FIGURE 14.4 Diagram of (a) cone pit and (b) long pit for outdoor storage of fruits and vegetables.

Vegetables and fruits (apples, pears) should not be stored in the same pit, although different kinds of fruits or vegetables can be stored in the same pits. Apples can be stored with pears, and potatoes with cabbage, and so on. Odors may be transferred from one product to another, and gases given off by fruits may affect the storage life of vegetables and vice versa. Generally, pits should be small. Once a cone-shaped pit is opened, all of the produce should be removed. Produce in long pits can be removed a part at a time, but protecting the opened end during cold weather may be a problem. Also, large pits are more difficult to open in cold weather. Having a mixture of different vegetables or fruits in a pit makes it unnecessary to open more than one or two pits at a time.

Pits should be placed on a different site every year to avoid contact of the newly stored produce with diseases that developed on the previously stored crop.

Barrels or other drums may be buried on top of the ground or shallowly by covering them with several layers of straw and soil. Growers should be sure that the barrels contained no previously stored materials that could harm the produce or the consumer. Barrel pits may be more convenient to open in cold weather than conical or long soil-covered pits.

MANAGEMENT OF STORAGE FACILITIES

Sanitation

Only sound fruits and vegetables should be placed in storage. Produce that shows signs of decay should be removed from storage. It may not be practical to remove spoiled produce from pits, so it is especially important that produce placed in pits be well suited for storage. All containers should be removed from basement, cellar, or outbuilding storerooms at least once a year and cleaned. The containers should be washed with household detergent and water or with dilute bleach (diluted to 10% of original strength) and air dried in the sun. The walls and ceilings of structures should be washed with the detergent or bleach solutions. Walls and ceilings can be whitewashed to sanitize the area and to provide further protection against diseases that may enter with or after placement of produce in storage. Aluminum-painted inside surfaces help in cleaning in the absence of whitewashing.

Temperature

Many fruits and vegetables store best at temperatures near 32°F. To maintain this temperature in the storeroom, outdoor temperatures must be well below freezing. Respiratory activity of the stored produce emits heat and causes temperatures to rise. In basements, cellars, and outbuildings, temperatures are regulated by opening and closing ventilators (windows, doors, or other openings). Daily management of the ventilators is needed to control storage temperatures. When outside temperatures are higher than those in the storeroom, the ventilators should be kept closed so that heat does not enter the storeroom. When outside temperatures are slightly or moderately lower than those in the storeroom, ventilators should be opened to allow for exchange of the cool outdoor air with the warm air of the storeroom. The ventilators should be closed when the outside temperatures are well below freezing. Overventilating a storage room in below-freezing weather gives a risk that stored produce may freeze. Freezing will ruin the storage capacity of most fruits and vegetables. Dried beans and peas and most root crops are not damaged by freezing, unless the temperatures are very much below freezing and of extended duration. Alternate freezing and thawing of these crops will cause their deterioration. Hard freezing of dried beans or peas often is practiced to kill weevils that may infest the seeds in storage. Approximate freezing temperatures and recommended storage temperatures of some fruits and vegetables are reported in Table 14.1. Strong winds accelerate rates of ventilation and raise the possibility of freezing of stored produce. Generally, in an insulated storage facility, outside temperatures can fall to below 10°F before temperatures in the storeroom will fall below 30°F. Intermittent ventilation of the storeroom will be required for outside temperatures in the range of about 10 to 20°F, and continuous ventilation will be required for outside temperatures in the range of about 20 to 32°F. Above 32°, ventilation is required only if the outside temperature is cooler than that inside.

TABLE 14.1
Temperatures at Which Produce Freezes and Recommended Storage Temperatures for Produce

	Temperature (°F)	
Product	**Freezing**	**Storage**
Warm Conditions		
Tomatoes	31	55 to 70
Peppers	31	45 to 50
Pumpkins	31	55
Squash	31	55
Sweet potatoes	30	55 to 60
Cool Conditions		
Potatoes	31	35 to 40
Dry beans	*	32 to 40
Dry peas	*	32 to 40
Root crops (carrots, beets, turnips, radishes)	*	32 to 40
Onions	31	32
Cold Conditions		
Cabbage	30	32
Celery	31	32
Apples	29	32
Pears	29	32
Citrus	30	32

* Produce can withstand freezing. Adapted from *Home and Garden Bulletin 119*. 1960. United States Department of Agriculture, Washington, DC.

Humidity

Except for dried produce and onions, stored fruits and vegetables will progressively lose water, shrivel, diminish in quality, and become unfit to eat without proper moisture in the storage environment. Dried beans and peas and onions should be stored in a dry, cool place.

Humidity can be maintained at a level suitable for most fruits and vegetables by adding water to the storage area or by placing the produce in vented plastic bags or plastic-lined boxes. Water can be sprinkled directly and frequently onto the floor of the storeroom. Placing and wetting sawdust or straw is a more effective method of increasing humidity than direct sprinkling of the floor. The straw and sawdust hold water and give a more even and prolonged control of humidity than sprinkling. Pans of water may be placed in the storeroom to help maintain high humidity. These practices are suitable for storage of fruits and vegetables other than root or tuber crops.

Maintaining moisture in root and tuber crops is difficult, and sprinkling, wetting of straw or sawdust, and placing of pans of water will not keep these crops from shriveling. Root or tuber crops may be buried in moistened straw, sawdust, or even soil in the storeroom. Plastic bags are excellent for storage of root or tuber crops. Bags can be tied, but tied bags should be slitted (1/2-inch cuts) in several places on the side for ventilation. Bags can be left open, but folded at the top, and placed in boxes. Root and tuber crops can be stored well also in plastic-lined boxes with loose coverings of plastic sheeting.

Waxing of vegetables is a possibility to prevent water loss, but waxing is not recommended for home storage because of the inconvenience of doing so.

HANDLING OF PRODUCE

All produce should be handled carefully during harvest and transport to storage. Punctured, cut, crushed, or otherwise damaged produce should not be stored. Containers for harvesting and storage should have smooth surfaces and should be inspected to ensure that no nails, staples, or other sharp objects protrude inwardly.

Fruits and vegetables should not be stored together, because of the possibility of transfer of flavors or odors and bad effects on storage of the mixed produce. Many vegetables can be stored together. The proper conditions—temperature, humidity, and ventilation—of storage must be met for each type, so some grouping of vegetables in storage according to the required or available conditions will be necessary.

Onions require some curing before storage. A few days in the sun after harvest usually provides sufficient curing. Onions have unique storage requirements that often require that they be stored alone. Onions should always be mature and dry for storage, and ones with thick or open necks should not be stored. Onions grown from seeds store better than those grown from sets, because of the narrower neck on the seed-grown onions. Containers in which onions are stored should be well ventilated and filled only about half. Mesh bags and slatted crates are excellent for onions. Humidity should be low for onion storage.

Potatoes also should be cured before storage, by holding them for a week or two in moderately warm (about 70°F) conditions. They should not be exposed to the sun or to drying winds during curing. The storage area for potatoes should be dark to prevent their greening. Greened potatoes should not be eaten because of the synthesis of a toxic alkaloid during greening. Potatoes in cool storage may accumulate sugars and taste sweet. Removing from storage and holding for a week at 70°F allows for metabolism of the sugars and restoration of the starchy taste.

Pumpkins and winter squash (except acorn squash) also require curing before storage. A week or 10 days at warm temperatures (about 80°F) accomplishes curing. If outdoor temperatures do not reach this level, place the pumpkins or squash in a warm room. Curing toughens the rind and heals cuts. Storage temperatures for pumpkins and squash have a narrow range around 55°F. These vegetables cannot stand chilling, and above 60°, they dry, and the flesh becomes stringy. Pumpkins and squash should be harvested with the stem attached. Removal of the stem provides an entry for pathogens.

Root crops (turnips, rutabagas, carrots, winter radishes) should not be stored until late fall. They keep well in the garden even with freezes. The dug crops should be cleaned and immediately placed in storage, as curing is not needed and desiccation could occur if storage is delayed. Shoots should be removed about 1/2 inch above the crown. If the roots are washed during cleaning, they should be dry before storing.

Tomatoes should be harvested without the stem of the fruit attached, even though some growers believe that this structure makes the fruit more attractive in marketing. Tomatoes and green peppers should be cleaned by washing, for wiping can be abrasive with attached soil, sand, or grit and cause damage. Tomatoes harvested in a mature-green stage will ripen in about 2 weeks in moderate warmth (about 70°F). Less mature tomatoes require more time to ripen or will not ripen. Cooling storage to 55 to 60°F slows ripening. To maintain quality, do not store tomatoes below 50° for more than a few days.

Cabbage requires no curing. For pit storing, cabbage can be pulled and stored upside down with roots attached. Heads of cabbage can be stored on shelves in outbuildings or cellars. Avoid storing of cabbage in basements because of the odors emitted from the cabbage.

Only late-maturing varieties of fruits or those purchased in the market during the winter should be considered for storage. High humidity and cold temperature are required to prevent their shriveling and ripening too much in storage.

Glossary

Active acidity: A term used to describe the soil acidity in a soil solution. It is the acidity that is measured by a common soil test and is expressed as pH.

Adsorption: The attraction of ions or compounds to the surfaces of solids.

Aggregate: Many soil particles held in a single mass or cluster to form a soil structural element, such as a sand-sized particle, gravel-sized particle, crumb, clod, block, or prism.

Ammonification: The biological process by which ammonium is released from nitrogen-containing organic matter. Ammonification is nitrogen mineralization.

Anaerobic: Without molecular oxygen.

Anion: A negatively charged ion.

Banding: Application of fertilizer in a concentrated linear zone in the soil, commonly alongside the rows of crops and under the soil surface. Contrast with **Broadcasting**, which is wide-zone application of fertilizer.

Bedding: Straw, woodchips, sawdust, paper, or other carbonaceous organic matter added to barn floors to absorb liquids from animal excrement and urine. Bedding increases the carbon:nitrogen ratio of farm manures.

Beneficial element: An element that may enhance yields, improve plant growth, or be required by a few plants, but not all plants. Cobalt, silicon, aluminum, sodium, selenium, and other elements.

Biomass: The amount of living material in an organism.

Buffer pH: A measurement of soil acidity, using a specific extract that is resistant or buffered to changes in pH. The value is applied in the calculation of soil lime requirement.

Bulk density: Mass of soil per unit volume of soil.

Calibration: When used in conjunction with soil tests refers to research that is performed to interpret soil tests and to correlate crop growth and yields to recommendations based on soil tests.

Carbonaceous matter: Organic matter with a high proportion of carbon and low proportions of other plant nutrients. Generally having a carbon:nitrogen ratio greater than 35:1.

Carbon:nitrogen ratio: The weight ratio of carbon to nitrogen in organic matter. Used as a term to project whether nitrogen will be mineralized from the organic matter and made available for plants to absorb or immobilized from the soil and made unavailable to plants.

Cash crop: A crop grown on land for income or for consumption by humans.

Catch crop: A crop that is grown to lessen leaching of nutrients during a period of time when a cash crop is not on the land.

Cation: A positively charged ion.

Chelate: In plant nutrition, a complex between an organic compound and a metallic plant nutrient, usually a micronutrient.

Chemical fertilizer: A common nondefinitive term used to identify manufactured fertilizers or some concentrated, soluble fertilizers of natural origin. Used in contrast to organic fertilizer.

Chlorosis: The loss of chlorophyll from plant leaves.

Clay: Soil separates less than 0.002 mm in effective diameter. Several types of clay minerals are in soils. Micaceous clays have expanding lattices, which swell and shrink when wetted and dried. Kaolinitic clays do not expand when wetted or shrink when dried. Clay is also a textural class of soil, generally >40% clay by mass.

Colloid: Organic and inorganic particles of very small size, for example, clay-sized particles, with large surface areas per unit of mass.

Complete fertilizer: A fertilizer that supplies nitrogen, phosphorus, and potassium.

Compost: Organic matter that is rotted before it is added to soil. Composting is employed to reduce the carbon:nitrogen ratio, kill plant diseases and weed seeds, and make the organic matter fit to apply to soil.

Contour: An imaginary line on the land or a line on a map connecting points of equal elevation on the surface of the land.

Covalent: A covalent bond is a chemical bond that is characterized by the sharing of pairs of electrons between atoms of a molecule.

Cover crop: A crop that is grown to stabilize soil against erosion during a period of time when a cash crop is not on the land. Sometimes used as a term for a **Green manure** crop.

Crop rotation: A planned sequence of crop production in a regularly occurring succession on land. Contrast with monoculture of a continuous crop or with random sequence of cropping.

Crust: A thin, dry, compact surface layer on soils.

Deflocculate: To cause soil particles to disperse. To bring about dispersion of clays from an aggregate.

Denitrification: The biological reduction of principally nitrate (or nitrite) to molecular nitrogen or gaseous nitrogen oxides. Denitrification is a major mechanism for losses of nitrogen from soils that undergo cycles of wetting and drying.

E horizon: A soil horizon between the A horizon and B horizon. Maximum eluviation has occurred in the E horizon, removing organic matter, iron oxides, clay, and other materials and causing the E horizon to be lighter colored than the A or B horizons. Formerly, the A2 horizon.

Earliness of soil: A characteristic of soils that drain and warm quickly for early planting in the spring.

Eluviation: The downward movement of solid material from a soil horizon (an A or E horizon). The deposition in the B horizon is called **Illuvation**. Movement may be by leaching.

Enzymes: Proteins that catalyze or increase the rate of chemical reactions in organisms.

Erosion: The wearing away of soil by geological agents (wind, water, ice).

Essential element: A plant nutrient. One of 17 elements required for plants to complete their life cycles: carbon, hydrogen, oxygen, nitrogen, phosphorus, potassium, calcium, magnesium, sulfur, iron, zinc, copper, manganese, molybdenum, nickel, boron, and chlorine. No other element fully substitutes for an essential element, and all plants require the essential elements. Having a beneficial effect on plant growth or development is not a characteristic of essentiality. Accumulation of an element in plant tissues is not a characteristic of essentiality. All plant nutrients are essential elements; thus, it is redundant to refer to nutrients as essential nutrients.

Exchange capacity: The total ions that a soil can hold or adsorb by electrostatic attraction. Also called *cation exchange capacity* to identify adsorption of cations by soil particles. Most of the exchange capacity of soils resides on the colloids, which are clays and organic matter.

Fallow: Cropland left idle to restore productivity through the accumulation of water, nutrients, or organic matter and through management of weeds.

Farm manure: The feces saved from farm animals and applied to land to provide plant nutrients.

Fertilizer: Material that is added to soil or directly to plants to supply plant nutrients. Fertilizers may be identified as being organic, chemical, or complete and having a certain analysis or grade.

Fertilizer grade (fertilizer analysis): The guaranteed minimum percentages of available nitrogen (N), phosphoric acid (P_2O_5), and potash (K_2O) in a commercial fertilizer.

Fine earth: The soil separates or sand, silt, and clay of soil.

Friable: A physical property describing a loose, crumbly soil.

Furrow slice: Also *acre furrow slice*. The layer of soil that is moved by tillage, usually by plowing. A furrow slice commonly is 6 inches deep. An acre furrow slice is the weight of the soil turned by tillage and is set at 2 million pounds. The weight of acre furrow slice is used in calculations of fertilizer applications, liming, and other soil amendments.

Green manure: A growing, immature crop that is incorporated into the soil to improve soil fertility.

Ground: The soil of the land.

Heavy metals: Metallic elements, some of which are plant nutrients but commonly are elements considered to be pollutants. These metals may have high densities, exceeding 4 or 5 g per cubic centimeter, and may include Cd, Co, Cr, Cu, Fe, Hg, Mn, Mo, Ni, Pb, Zn, and perhaps others, not considered as metals, such as As. Elements have metallic properties at room temperature. List may include elements between copper and bismuth on the Periodic Table of Elements.

Horizon: See **Soil horizon**.

Humus: Dark-colored (black or brown), somewhat stable form of soil organic matter remaining after relatively easily decomposable plant and animal residues have degraded.

Illuviation: The deposition, normally in a B horizon, of materials moved downward by eluviation.

Immobilization: The process by which available plant nutrients in soil are rendered unavailable by microbial consumption. Immobilization usually following incorporation of carbonaceous materials into soils.

Ion: An atom, group of atoms, or compound that is electrically charged. An anion (–) or a cation (+).

Land: In agriculture, the natural environmental, nonwater, area of the earth in which crops are grown.

Leaching: In soils, the downward movement of materials that dissolve in water as the water passes through the soil. See **Eluviation** and **Illuviation**.

Legume: A plant of the family Fabaceae (Leguminosae), usually associated with nitrogen fixation.

Lime: A material composed of carbonates, oxides, or hydroxides of calcium, magnesium, or both elements and used to neutralize soil acidity. Commonly, agricultural limestone or mixtures of calcium and magnesium carbonates.

Lime requirement: The amount of lime that is required to raise the pH of an area of land or mass of soil to a desired value, usually between pH 6 and 7. The limestone required is that needed to correct the active acidity and reserve acidity.

Limiting factor: The minimum amount of a nutrient or condition that is needed for plant growth. Liebig's Law of the Minimum states that the growth of a plant cannot exceed that which is supported by the limiting factor.

Loam: A medium-textured soil.

Lodging: Falling over of plants. Irreversible displacement of a plant from its upright position.

Luxury consumption: The absorption by plants of more nutrients than the plant needs, but without the expression of any symptoms of toxicity from the nutrients.

Macronutrient: A plant nutrient that is present in plant biomass in relatively high concentrations. Carbon, hydrogen, oxygen, nitrogen, phosphorus, potassium, calcium, magnesium, and sulfur. The latter six are called *soil-derived macronutrients*.

Mechanical analysis: A process by which soil separates, sand, silt, and clay, are measured quantitatively by mass.

Micronutrient: A plant nutrient that constitutes a low proportion of plant biomass. Also known as *trace element* or *minor element*. Iron, zinc, copper, manganese, molybdenum, nickel, boron, and chlorine.

Mineralization: The process by which plant nutrients in soil organic matter are released by microbial action into soluble or available forms, which plants can absorb. Nitrogen mineralization is ammonification.

Mineral soil: A soil dominated by inorganic matter, primarily sand, silt, and clay, and usually containing less than 20% organic matter by mass.

Mulch: A layer of material applied across the surface of the ground.

Necrosis: Death associated with tissues of leaves and other plant organs.

Neutral soil: A soil for which the surface area (A or Ap horizon) is not acidic or alkaline. Practically, a soil with pH between 6.6 to 7.3.

Nitrification: The oxidation of ammonium to nitrate in soils or other media. Process is microbiologically mediated.

Nitrogen fixation: A biological process by which gaseous nitrogen in the atmosphere is converted into ammonium. In agronomic practice, the process is symbiotic between legumes and bacteria, which colonize roots of the legumes.

Nitrogenous material: Organic matter that is relatively high in nitrogen in relation to carbon with C:N ratios of <35:1.

Nonexchangeable cations: Positively charged ions that are held in the lattices of micaceous clays. Also referred to as fixed ions, such as fixed potassium.

Nonlegume: A crop that is not a legume. Usually used as a term to identify plants that do not perform nitrogen fixation.

Organic fertilizer: A fertilizer of natural mineral or biological origin, usually processed only physically not involving chemical manufacturing or alteration. Identification may be provided as a permitted use in listings by a certifying organization. Term is used in contrast to chemical fertilizer. Other criteria may include considerations of nutrient concentration and solubility in the naturally occurring material.

Organic soil: A soil that has a high percentage of organic matter, commonly set at >20% by mass, throughout the A and B horizons.

Pan: A layer of soil that is compacted, hardened, or high in clay. Hardpan, claypan, fragipan.

Ped: A natural structural unit in soil.

pH: An expression of hydrogen ion concentration in soil or a solution. The negative logarithm of hydrogen ion molar concentration.

Phosphorus fixation: A chemical process by which phosphate in solution is precipitated from solution and into relatively insoluble iron and aluminum compounds in acid soils and into sparing soluble calcium and magnesium compounds in basic soils.

Plant nutrient: An essential element.

Potassium fixation: A physical process by which potassium ions are trapped in the lattice of micaceous clays.

Primary mineral: A mineral that has not been altered chemically. Contrasted with secondary mineral (clays).

Puddled soil: Dense, massive, structureless (no aggregates) soil artificially compacted when wet. Characteristic of structureless compressed clay. Derived from the definition of the verb *to puddle*, to make muddy.

Reserve acidity: Soil acidity that is held to the soil particles. Reserve acidity is the major fraction of acidity in soils.

Runoff: Also *surface runoff*. The portion of rainfall or irrigation that moves across the surface of the soil without entering the soil.

Salinity: The amount of soluble salts in a soil or dissolved in a soil solution or nutrient solution.

Sand: A soil separate with effective diameters from 0.05 to 2 mm. Sand is also a soil textural class in which sand is the dominant separate.

Secondary mineral: A mineral, such as clay, formed from decomposition of primary minerals and reconstitution into a new mineral.

Seedbed: Soil or land prepared to promote germination of seed or receipt of transplanted seedlings and to support their subsequent growth.

Sidedress: To apply fertilizer alongside a crop growing in a row. Application is on the soil surface or just under the surface.

Silt: A soil separate with effective diameters from 0.002 to 0.05 mm. Silt is also a soil textural class in which silt is a principal separate.

Slag: A product of smelting. Basic slag is a by-product of iron smelting and is sometimes used as a phosphorus-containing fertilizer.

Soil: The collection of surface matter of land that supports plants and that has properties affected by parent material, climate, vegetation, topography, and time and the interactions of these factors.

Soil acidity: The intensity of hydrogen ion concentration in soils. The total acidity is the sum of active acidity and reserve acidity.

Soil aeration: The movement of air from the atmosphere into the soil pores. The movement of air through soil pores.

Soil air: The gaseous phase of soil; the volume of soil not occupied by solid or water.

Soil amendment: A material added to soils to change their chemical or physical properties. Often considered not to be added to soil for the purpose of supplying plant nutrients.

Soil chemical properties: Characteristics of soil that are defined by chemistry, such as the composition and reaction of soil constituents. Soil pH and nutrient supply are chemical properties contributing to soil fertility.

Soil color: A property of soil distinguished by the light that it reflects. Color has the components of *chroma* (strength, purity, saturation), *hue* (gradation), and *value* (lightness, darkness).

Soil drainage: The percolation of water through a soil. A well-drained soil is free of saturation and has considerable pore space filled with air.

Soil fertility: An expression of the capacity of soil to support crop production. Fertility is based on soil chemical, physical, and biological properties. Often soil fertility is defined as the capacity of soils to provide nutrients to plants. Synonyms are *soil quality* and *soil health*.

Soil health: See **Soil fertility**.

Soil horizon: A layer of soil lying about parallel with the soil surface. Characteristics of horizons are used in soil classification or naming. The horizons of a soil are referred to collectively as a soil profile. The most common soil horizons are the A, B, E, and C horizons. These horizons are named according to their position in the profile and degrees of weathering and leaching.

Soil organic matter: The carbon-containing constituents derived from formerly living organisms. Humus is a dark-colored, stable form of soil organic matter.

Soil physical properties: Characteristics of soil that are defined by matter and energy. Physical properties include such characteristics as tilth, structure, drainage, water-holding capacity, aeration, and bulk density, which contribute to soil fertility.

Soil porosity: The portion of the soil that is filled with air or water.

Soil profile: A vertical section of soil, showing the horizons.

Soil quality: See **Soil fertility**.

Soil reaction: Soil acidity. Soil pH.

Soil separates: The sand, silt, and clay of soil, particles which will pass through a 2-mm sieve. Also called *fine earth*. Gravel, cobbles, flagstones, and other constituents larger than 2 mm in diameter are not soil separates but are used as modifiers in names determining soil texture, for example, gravelly fine sandy loam.

Soil solution: The aqueous liquid portion and its solutes in soil.

Soil structure: A soil physical property that is based on the arrangement of soil separates into groups of particles that adhere by cementation or cohere. Aggregates are clusters of many soil particles in a single mass. Ideally aggregates are sand-sized for good soil tilth. Large aggregates are called *clods*, *crumbs*, *block*, *prisms*, or *peds*.

Soil surface area: The total surface area of the sand, silt, and clay particles in soil.

Soil survey: The examination, description, classification, and mapping of soils of an area.

Soil testing: An analytical process applied to assess the capacity of soils to supply plant nutrients. Often a rapid, semiquantitative quick test to assess the availability of plant nutrients.

Soil texture: A soil physical property that is determined by the relative proportions by weight of sand, silt, and clay. Organic matter does not affect soil texture but may modify soil structure. Sands dominate in coarse-textured soils, which are referred to often as *light soils*. Fine-textured soils, called *heavy soils*, have considerable portions of clays and silt. Loamy soils are medium-textured soils. Soil textural class is part of a name of a soil and refers generally to the A horizon.

Soil tilth: The physical condition of soil (structure) in relation to crop growth.

Soil water-holding capacity: The water that is held in the fine pores and around the soil particles after water has drained from the large pores. *Field capacity* is a term for water-holding capacity in a soil, once saturated with water and drained for 2 or 3 days. The *wilting percentage* is the soil water content at which plants in soil permanently wilt. *Available water* is the soil water readily absorbed by plant roots and is the portion of soil water between the wilting percentage and field capacity. *Waterlogged* refers to soil saturated with water, with pores filled with water.

Strip cropping: The practice of growing crops in alternate zone of crops and sod.

Subsoil: The layer of soil under the topsoil, commonly referred to, but not specifically, as the *B horizon*.

Tillage: Mechanical manipulation of soil to modify soil for crop production. Also called *cultivation*. *Conservation tillage* is a practice of tillage to reduce the loss of soil or water relative to conventional tillage commonly used to prepare a seedbed.

Tilth: See **Soil tilth**.

Tissue testing: An analytical process for assessing the nutritional status of plants by analysis of their tissues. Often a quick test for rapid, on-site assessment and recommendation for needs of fertilization.

Topsoil: Surface layer of soil moved by tillage. Commonly, but not specifically, the A horizon.

Volatilization: The escape of gaseous materials from a medium to the atmosphere. Ammonia volatilization is a principal mechanism for loss of nitrogen from alkaline soils and composts.

Weathering: The physical and chemical breakdown occurring in rocks and minerals at or near the surface of the earth or in the soil.

Bibliography

Adams, F. (ed.) 1984. *Soil Acidity and Liming*, 2nd Ed. Agronomy 12, American Society of Agronomy, Madison, Wis.

Agrios, G.N. 2005. *Plant Pathology*. Elsevier Academic Press, Amsterdam.

Barker, A.V. and D.J. Pilbeam (eds.). 2007. *Handbook of Plant Nutrition*. CRC/Taylor & Francis, Boca Raton, FL.

Brady, N.C. 1974. *The Nature and Properties of Soils*, 8th Ed. Macmillan, New York.

Brady, N.C. and R.R. Weil. 2002. *The Nature and Properties of Soils*, 13th Ed. Prentice Hall, Upper Saddle River, N.J.

Chapman, H.D. (ed.).1966. *Diagnostic Criteria for Plants and Soils*. University of California, Division of Agricultural Sciences, Berkeley.

Datnoff, L.E., W.H. Elmer, and D.M. Huber. 2007. *Mineral Nutrition and Plant Disease*. The American Phytopathological Society, St. Paul, Minn.

DeBach, P. (ed.). 1964. *Biological Control of Insect Pests and Weeds*. Reinhold, New York.

Department of Food Science, Washington State University. Reprinted 2008. *Storing Fruits and Vegetables at Home*. EB 1326, Washington State University Extension, Pullman, Wash.

Ellis, B.W. and F.M. Bradley. 1992. The *Organic Gardener's Handbook of Natural Insect and Disease Control*. Rodale Press, Emmaus, Pa.

Epstein, E. and A.J. Bloom. 2005. *Mineral Nutrition of Plants: Principles and Perspectives*, 2nd Ed. Sinauer, Sunderland, Mass.

Havlin, J.L. J.D. Beaton, S.L. Tisdale, and W.L. Nelson. 2005. *Soil Fertility and Fertilizers*, 7th Ed. Pearson Prentice Hall, Upper Saddle River, N.J.

Hoitink, H.A.J. and H.M. Keener (eds.). 1993. *Science and Engineering of Composting: Design, Environmental, Microbiological and Utilization Aspects*. Renaissance Publications, Worthington, Ohio.

Howard, A. 1943. *An Agricultural Testament*. Oxford University Press, New York.

Kristiansen, P., A. Taji, and J. Reganold. 2006. *Organic Agriculture. A Global Perspective*. Comstock, Cornell University Press, Ithaca, N.Y.

Lampkin, N. 1990. *Organic farming*. Farming Press, Ipswich, U.K.

Lotter, D.W. 2003. Organic agriculture. *Journal of Sustainable Agriculture* 21(4):59–128.

Mackay, S. 1979. *Home Storage of Fruits and Vegetables*. NRAES-7, Northeast Regional Agricultural Engineering Service, Cornell University, Ithaca, N.Y.

Marschner, H. 1995. *Mineral Nutrition of Higher Plants*. Academic Press, London.

Maynard, D.N. and G.J. Hochmuth. 2007. *Knott's Handbook for Vegetable Growers*, 5th Ed., Wiley, New York.

Metcalf, D.S. and D.M. Elkins. 1980. *Crop Production: Principles and Practices*, 4th Ed. Macmillan, New York.

Mills, H.A. and J.B. Jones, Jr. 1996. *Plant Analysis Handbook II. MicroMacro*, Athens, Ga.

Northbourne, Lord C.J. 1940. *Look to the Land*. J.M. Dent, London. [Perennis, S. 2003. 2nd Ed., revised, Hillsdale, N.Y.]

Powers, J.F. and W.A. Dick. (eds.) 2000. *Land Application of Agricultural, Industrial, and Municipal By-Products*. SSSA Book Series 6, Soil Science Society of America, Madison, Wis.

Rechcigl, J.E. (ed.). 1995. *Soil Amendments and Environmental Quality*. CRC/Lewis, Boca Raton, Fla.

Reed, H.S. 1942. *A Short History of the Plant Sciences*. Chronica Botanica, Waltham, Mass.

Russell, E.J. 1961. *Soil Conditions and Plant Growth*, 9th Ed. Wiley, New York.

Stevenson, F.J. (ed.). 1982. Nitrogen in Agricultural Soils. *Agronomy* 22, American Society of Agronomy, Madison, Wis.

Stoffella, P.J. and B.A. Kahn. 2001. *Compost Utilization in Horticultural Cropping Systems*. Lewis, Boca Raton, Fla.

Tate, R.L., III. 1987. *Soil Organic Matter: Biological and Ecological Effects*. Wiley, New York.

United States Department of Agriculture. 1938. Soils and Men, *1938 Year of Agriculture*, United States Department of Agriculture, Washington, D.C.

United States Department of Agriculture. 1966. Storing Vegetables and Fruits in Basements, Cellars, Outbuildings, and Pits. *U.S. Department of Agriculture Home and Garden Bulletin 119*, Market Quality Research Division, Agricultural Research Service, Washington, D.C.

United States Environmental Protection Agency. 2009. *Wastes—Resource Conservation—Reduce, Reuse, Recycle—Composting*. http://www.epa.gov/osw/conserve/rrr/composting/index.htm. Accessed July 21, 2009.

Waksman, S. 1936. *Humus*. Williams & Wilkins, Baltimore.

Ware, G.W. and J.P. McCollum. 1968. *Producing Vegetable Crops*. Interstate, Danville, Ill.

Westcott, C. 1964. *The Gardener's Bug Book*. Doubleday, Garden City, N.Y.

Index

N

Q

R

S

T

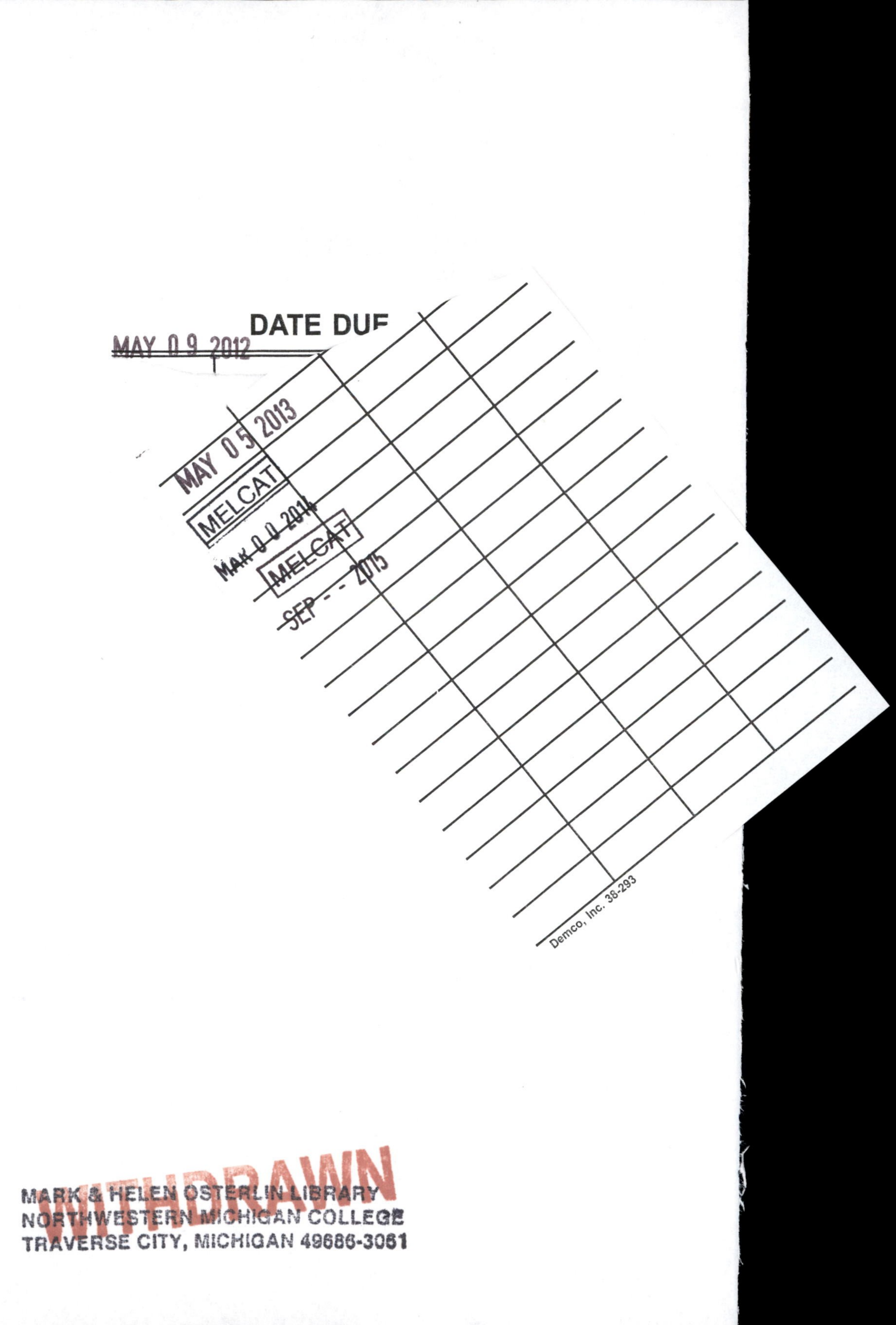

DATE DUE
MAY 0 9 2012
MAY 0 5 2013
MELCAT
MAR 0 8 2014
MELCAT
SEP - - 2015
Demco, Inc. 38-293